KU-450-196

WITHDRAWN
FROM STOCK

The Life Puzzle

On Crystals and Organisms and on the Possibility of a Crystal as an Ancestor

A. G. CAIRNS-SMITH
Lecturer in Chemistry,
University of Glasgow

The Life Puzzle

On Crystals and Organisms and on the Possibility of a Crystal as an Ancestor

OLIVER AND BOYD
EDINBURGH

First published 1971 by
OLIVER & BOYD
Tweeddale Court, 14 High Street, Edinburgh EHI IYL
A Division of Longman Group Limited

ISBN 0 05 002297 0

Printed in Great Britain by
Bell & Bain Ltd., Glasgow

Preface

'A grey rock, said Ruskin, is a good sitter. That is one type of behaviour. A darting dragon-fly is another type of behaviour. We call one alive and the other not. But . . . to make "life" a distinction between them is at root to treat them both artificially'.
SIR CHARLES SHERRINGTON, 1937 *Gifford Lectures.*

This book is concerned with the connections between 'non-living' and 'living' forms of the organisation of matter.

Part I is mainly descriptive. It is about crystals and organisms, and also about objects whose organisation is related to both—the nucleic acids and proteins.

Part II is concerned with theoretical connections between crystals and organisms. 'How in general can atoms organise themselves?' 'What particular features of the resulting arrangements should we expect?' 'What fundamental difference is there, if any, between a rock and a dragon-fly?'

Part III is on the origin of life. Here is the most direct connection between 'non-living' and 'living'. I propose a somewhat heterodox view: that the origin of life on earth was more closely associated with crystals—with rather insoluble inorganic crystals—than with the organic polymers that underlie the workings of all the organisms that we know today. I suggest that the currently orthodox idea of the crystal as the archetype of 'non-living' and the organism of 'living' is off the point: that a crystal is indeed conceivable as an *ancestor*.

GRAHAM CAIRNS-SMITH

Acknowledgements

My thanks are due to my wife and to friends and colleagues for their (usually) constructive criticism.

I would like to thank also Dr A. Keller for providing Plates I and II, Dr E. A. C. Follett for Plate III, Dr A. H. Weir for Plate VI, Mme Agnès Oberlin for Plates VIII and IX, Mr Donald W. MacKenzie for Plate X, Dr M. F. Perutz for Fig. 22 and Dr Thomas Baird for making Plates IV, VII and VIII. I am grateful also to Dr Keller for permission to reproduce Fig. 17 and to Professor Waddington for permission to reproduce Figs. 35 and 36.

I acknowledge also the kind permission of *Kolloid-Zeitschrift & Zeitschrift für Polymere* (Plates I and II), Pergamon Press (Plate VI), Butterworths (Fig. 17), George Allen and Unwin Ltd. (Figs. 35 and 36) and Academic Press Inc. (Figs. 12, 13, 15, 45, 46, 47 and 48) for permission to reproduce illustrations and of Cambridge University Press for permission to reproduce two passages from Sherrington's *Man on his Nature*.

Contents

1

Introduction

Archimedes reckoned that 'a sphere of the size attributed by Aristarchus to the sphere of the fixed stars would contain a number of grains of sand less than 10 000 000 units of the eighth order of numbers' (Heath, 1897). In modern notation this is 10^{63}. Archimedes had discovered one of the great wonders of arithmetic: that one can express unimaginably large numbers quite compactly.

A more recent example of a vast number which can be easily written appears in Eddington's estimate that the total number of electrons in the universe is about 10^{79}.* We may now think that this was too small a view of the universe, but around 10^{80} is still as big a number of any real objects that we are ever likely to want to talk about. Yet it is by no means the largest number that some quite sensible people do talk about. Archimedes himself was not content with numbers like 10^{63}: he went on to consider 'a myriad-myriad units of the myriad-myriad-th order of the myriad-myriad-th period'—that is to say $(10^{8\times 10^8})^{10^8}$.

A WORLD IN A GRAIN OF SAND

The *really* big numbers become important when we come to consider not simply how many units there are in a given region, but how many ways they can be arranged. An example can be found in elementary organic chemistry. A class of hydrocarbons can be formed by linking together carbon and hydrogen atoms in such a way that every carbon atom is joined to four other atoms and every hydrogen atom to just one. Excluding ring systems, members of this class have the general formula C_nH_{2n+2}. (For example, $n = 1$ corresponds to methane, CH_4;

---

* Eddington, 1935, supposed that the universe contains about a million million (10^{12}) galaxies and that each galaxy contains about ten thousand million (10^{10}) sun masses. The mass of the sun is about 10^{27} tons, i.e. 10^{33} grammes, and a gramme of matter contains about 10^{24} electrons: $10^{12}\times 10^{10}\times 10^{33}\times 10^{24} = 10^{79}$.

$n = 2$ to ethane, C_2H_6; and so on.) The following arrangements are possible for $n = 1$ up to $n = 4$:

```
    H            H  H             H  H  H
    |            |  |             |  |  |
  H-C-H        H-C--C-H         H-C--C--C-H
    |            |  |             |  |  |
    H            H  H             H  H  H

   n=1           n=2               n=3

   H  H  H  H                 H  H  H
   |  |  |  |                 |  |  |
 H-C--C--C--C-H      or     H-C--C--C-H
   |  |  |  |                 |  |  |
   H  H  H  H                 H  |  H
                                 |
       n=4                     H-C-H
                                 |
                                 H

                                n=4
```

Notice that when you reach 4 carbon atoms there are two possible arrangements (isomers). It can easily be checked that with 5 carbon atoms there are 3 isomers, with 6 carbon atoms there are 5 isomers, and that the number of possibilities rapidly increases as the number of carbon atoms is further increased. You may like to check that there are 35 ways of arranging 9 carbon atoms (with their 20 hydrogens), but you will probably be prepared to take Henze and Blair's word for it that there are 61,491,178,805,831 isomers of $C_{40}H_{82}$ (Henze and Blair, 1931). At this rate it is clear enough that 200 carbon atoms could be arranged, with 402 hydrogens, in far more than 10^{79} different ways. Yet one such molecule of $C_{200}H_{402}$ could only just be seen with the most recently developed electron microscopes. There is a sense, then, in which a submicroscopic speck of matter, incomparably smaller than any one of Archimedes' sand grains, can encompass a number greater than the number of things in the observable universe. There are far more possible $C_{200}H_{402}$'s, that is, than there are actual electrons in the observable universe. And this is in spite of the quite restrictive rules that govern the construction of this class of hydrocarbon molecules.

Science, you may say, should not concern itself with such airy speculations. Does it really matter how many ways the atoms in a system *might* be arranged? Should we not concentrate on trying to determine how they actually *are* arranged in particular instances?

We may indeed be interested in actual arrangements of atoms: in crystals, for example, or in the molecular components of organisms. But we are interested also in the number of ways in which the atoms are *not* arranged—but might have been. Such knowledge may help us in making a broad estimate of the *organisation* of a particular arrangement.

We will assert that to organise is to select, to cut down possibilities. Crystals and organisms are particularly vivid examples of the capacity of matter to organise itself: to select more or less special arrangements of atoms out of Archimedean numbers of possibilities. To estimate 'the amount of selection' involved in particular cases we have to know just how Archimedian the possibilities are.

GASES AND GOLF

Let us consider for a moment the classical example of a most disorganised state of matter—an 'ideal' gas. This consists of a hectic, thinly dispersed crowd of rapidly and randomly moving particles continually colliding with each other. The collisions are always elastic, as are collisions with the sides of any containing vessel, so the system can remain in this highly dynamic state indefinitely. At higher temperatures the molecules of a given gas move about faster, on average, than at lower temperatures. It can be shown that this model accounts in some detail for several important properties of gases. For example, a gas will expand spontaneously, and continue to do so until restrained by some kind of container. The random motion of the particles is enough to explain their tendency to move away from each other—there is no need to invoke a specific repulsive force. This may seem obvious enough when we consider such a particular example, yet this example illustrates an important principle which is less obvious when thought of generally: that a large assemblage of individuals may behave predictably, even although each individual separately is acting at random. And it is easy in principle to predict how such an assemblage will behave: it will move into the most disorganised state accessible to it, i.e. into a general state which allows the maximum possible freedom to the individuals.

Let us consider now the game of golf as an illustration of an organising procedure.

The main idea in golf is to hit a ball (B) in such a way that it gets nearer to a hole (H). Bad golfers are aware that this is a problem, because there are always more ways of ending up farther away from the hole than nearer (see Fig. 1). So to hit at random will usually make things worse. Indeed, your hitting must be distinctly better than random to make consistent progress.

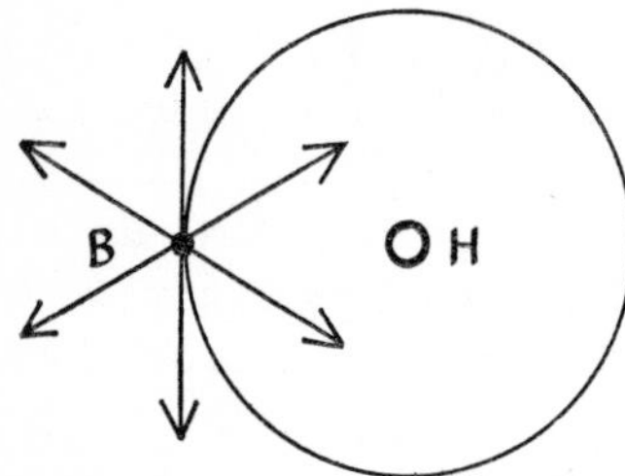

Fig. 1. Even for a short stroke, less than BH, there are more directions that will take your golf ball further away from the hole than nearer to it.

Similarly, an uncontained gas expands because there are always more directions 'outward' than 'inward', and so a greater chance that any of its randomly moving particles will go that way. (Put differently, there are more ways of being 'expanded' than 'contracted', so the shuffling effect of molecular motion will be more likely to shift the system into the expanded state. When one is dealing with a large number of molecules—say 10^{23}—the scope for being expanded is greater to an Archimedian degree, so left alone the self-shuffling system is virtually certain to choose the expanded state.)

Now, to hit a golf ball onto a green is to organise: to select any of a small number of positions out of a much larger number of possible positions. As we have seen, both in gases and golf the tendency is for random processes to shift a situation from one in which there is a relatively small number of possibilities into one in which there is a larger number. There are, for example, more positions for the ball conforming to the general description 'on the fairway' than there are for the general description 'on the green'. It is for this reason that the random golfer will tend to shift his ball off the green onto the fairway; and indeed, given the chance, right off the golf course. The tendency is, then, for random processes to 'de-select', to disorganise. And this tendency is quite general in considering the behaviour of atoms and molecules. Since they are subject to random motion there is a continual tendency for multi-particle systems to become increasingly disorganised: to move from a general state in which there is a relatively small number of possibilities, into a general state in which there is a larger number of possible arrangements for the parts.

MICRO-STATES AND MACRO-STATES

By an 'arrangement' of molecules in a gas we may mean simply their arrangement in space. More generally, however, we are interested not only in positions but—perhaps even more—in how fast the molecules are moving, how violently they are vibrating, and so on. We can extend the idea of an 'arrangement' to include a specification not only of the positions but also of the motions of the individual molecules. The term *micro-state* is preferable in this case. To define the micro-state of, say, the air in a room at a given instant would involve a separate description of every molecule. This is, needless to say, quite impracticable. Apart from anything else, the description would require far more paper and ink than has ever existed. By contrast the *macro-state* of a system is its state when considered from a large-scale point of view. For the air in a room it would represent the pressure, temperature, volume, etc. We can describe macro-states easily enough, but from a molecular point of view such descriptions are almost unimaginably vague: a system in a given macro-state, e.g. 10 000 cu. ft of air at 762 mm pressure and 18·63°C, can exist in an Archimedian number of micro-states: indeed, so many are the possibilities for the air in a room that there is virtually no chance that it would ever be in the same micro-state twice, even though the molecules may be changing between micro-states more than 10^{40} times every second—and even if you kept it for a period greater than the age of the Earth. (You could multiply this period by 'the number of electrons in the universe' if you like, but it would not help very much.)

Yet such numbers—numbers of possible micro-states underlying a given macro-state—are to be taken seriously. They are not to be dismissed as 'just infinite'. They are of crucial importance, or at least their order of magnitude is of crucial importance, in determining the general behaviour of physico-chemical systems: they are measures of the necessary vagueness of macro-descriptions, of how disorganised macro-states are.

Moreover it is possible to calculate roughly numbers of micro-states for simple systems. It turns out that 2 g of hydrogen at 25°C and atmospheric pressure can exist in about $10^{4\,100\,000\,000\,000\,000\,000\,000\,000}$ micro-states, i.e. $10^{41 \times 10^{23}}$. For the same number of molecules of water vapour (i.e. 18 g) the figure is about $10^{59 \times 10^{23}}$; for liquid water it is about $10^{22 \times 10^{23}}$.

Suppose you have 18 g water vapour in a container at 25°C and atmospheric pressure. The molecules are moving about at random. So long as, at a given instant, the molecules are in any one of the $10^{59 \times 10^{23}}$ micro-states consistent with the general state 'vapour', no change will be apparent. But there is a chance that the molecules will happen on one of the $10^{22 \times 10^{23}}$ arrangements of energies and positions consistent with the liquid state. That is, just by chance, the vapour may suddenly turn into liquid. This chance is given by the ratio of the number of ways in which the molecules can be arranged in the liquid to the number of their possible arrangements in the vapour, i.e. $10^{22 \times 10^{23}} : 10^{59 \times 10^{23}}$, or one chance in $10^{37 \times 10^{23}}$; for all practical purposes—nil. Thus, in spite of the vast number of ways in which the molecules can 'be a liquid', the number of ways in which they can 'be a vapour' is so much vaster that the liquid would seem a wildly improbable state. Using the golf analogy again, you might be persuaded that if the greens on a golf course are 200 yards across then even a random golfer will land on one quite often. But not if the golf course is vastly bigger than the known universe.

By no stretch of credulity, then, can we explain the very common occurrence of liquid water in terms of the random motions of molecules. That matter exists in condensed, i.e. non-gaseous, states is evidence of the existence of organising factors which somehow persuade molecules to jostle themselves into relatively 'rare' arrangements. (The greens may be relatively very minute, but what happens if the golf course is shaped so that balls tend to roll onto them?)

FORCES OF ORDER

The organising factors accounting for condensed forms of matter are, of course, interatomic and intermolecular attractive forces. These are, very briefly:

1. *Covalent bonds* that hold atoms together in molecules. An atom, for example a carbon atom, consists of a minute intensely positively charged nucleus surrounded by a cloud of negatively charged electrons. When atoms are covalently bonded together, these clouds interpenetrate and to some extent lose their separate identity. A

covalent arrangement of atoms may remain intact for thousands of millions of years (Eglinton and Calvin, 1967). This faithfulness is the outstanding feature of covalent bonds: it makes possible the elaborate, durable, submicroscopic machinery of life. It also accounts for the durability of many other things—like diamonds and stones.

2. *Ionic bonds* operate between oppositely charged atoms and molecules, i.e. ions. For example, a crystal of common salt is a regular array of positively charged sodium and negatively charged chorine ions. These attractive forces are strong but not usually as faithful as covalent bonds can be.

3. *Bonds in metals* are obviously often very strong: in a metal crystal each atom shares one or more of its electrons with all the other atoms. We will not have much to say about these forces.

4. *Van der Waals forces* are a rather varied collection of much weaker effects. They arise from two main considerations.

(*i*) Electrons in atoms and molecules are mobile. An approaching electric charge can alter their average distribution so as to create a *dipole*, that is a situation where the centre of negative charge due to the electrons does not coincide with the centre of positive charge due to the atomic nuclei. Such an induced dipole will be oriented so as to give an attractive electrostatic force (this is rather like a magnet attracting an unmagnetised nail). Even in the absence of an external charge the moving electrons will be continually creating momentary dipoles due to momentary asymetries in their distribution. These dipoles can induce momentary opposite (attractive) dipoles in adjacent molecules. Such *dispersion forces* are very weak, very short range but universal; they operate between all kinds of atoms and molecules.

(*ii*) Molecules often contain permanent dipoles so that they tend to stick together, to ions and to molecules in which they can induce a dipole.

We will not pursue much further here the detailed character of the intermolecular forces. For our purposes it is important to know that they exist, and that they depend for their strength and range on subtle features of the electronic structures of the molecules concerned. They are of great importance in determining the ways in which molecules pack in crystals and in determining the forms and functions of protein molecules. But it is still difficult to predict in detail their effects.

5. *Hydrogen bonds* are a particular kind of dipole interaction, but on account of their strength and directional character they require special mention. In some respects they are like weak covalent bonds. Molecules, like water, that contain the O—H group of atoms readily associate due to the strong permanent displacement of electrons towards the oxygen atom on such bonds. Water molecules, for example, tend to stick together like this:

```
H—O····H—O
  |      |
  H      H
```

Molecules that contain the N—H group are also active in forming hydrogen bonds with other molecules that contain oxygen or nitrogen atoms:

```
\          /            \          /
 N—H·····O       or      N—H·····N—.
/          \            /          \
```

(We will come to the particular importance of these interactions when we discuss the proteins and nucleic acids.)

ORDER V. CHAOS

There is, then, a conflict between thermal motion, tending to make atoms fly apart and distribute themselves as thinly as possible throughout available space—'the entropy effect'—and the various attractive forces which, operating by themselves, would make atoms pack closely together—known as 'the energy effect' (because such giving way to attractive forces results in a lowering of potential energy). This conflict is crucial in considering the directions of change in physico-chemical systems. A system which is isolated from environmental disturbance will tend to change in such a way as to create a balance between the two tendencies—'the forces of order' and the chaotic thermal shuffling. On the one hand, the system is tending to reach a condition of lowest possible potential energy, on the other it would like to be in a condition consistent with the maximum possible number of micro-states: this means the highest possible *entropy*.

The entropy, S, of a system is related to the number of microstates, m, thus:

$$S = k \log m,$$

where k is Boltzmann's constant (about $0{\cdot}33 \times 10^{-23}$ cal/deg). So the entropy for 18 g water at 25°C and atmospheric pressure is about 45 cal/deg for the vapour, and about 17 cal/deg for the liquid. Differences between entropies correspond to ratios of (Archimedian) numbers of microstates since:

$$S_2 - S_1 = k \log \frac{m_2}{m_1}$$

OPEN SYSTEMS

In general, systems left to themselves tend towards *equilibrium*, that is a state of balance between energy and entropy effects. But if systems are not left to themselves, if they are continually being disturbed, then they may remain off-balance indefinitely. Rivers, flames, whirlpools, are examples of systems which are maintained in an off-balance condition through continual 'interference' by the environment. Such systems, in which matter and energy are being exchanged with the environment, are called *open systems.* If their form remains constant, like a sheltered candle flame, then we describe the condition as a *steady state.*

The boundary between an open system and the rest of the universe is necessarily rather arbitrary. You may think of an entire river as an open system. But so is any given stretch of a river, or a whirlpool within a particular stretch, or an eddy in the whirlpool, or a sub-eddy within that. Or, going in the other direction, you may think of a river as part of a larger open system constituting the water cycle: of the water cycle, even, as part of the standing turbulence generated continuously by the flux of photons from the sun. Even this flow can be thought of as part of a broader cosmic flow in which the Sun itself is in a temporary state of inbalance. This is the Heraclitean analogy of the universe as a river. The stream is far from smooth, consisting of a complex network of open systems within open systems. If we accept the general cosmology of Gamow (1961), the cosmic flow is the expansion of the universe from an initially supercondensed state; and the first major hierarchy of eddies constituted the break-up of the initial material into galaxies and then stars. But the most subtle, if not the biggest, of the sub-eddy systems in the cosmic flow are the organisms.

Sherrington in 1937 thought of organisms in this sort of way: 'The living energy-system, in commerce with its surround, tends to increase itself. . . . If we think of it as an eddy in the stream of energy it is an eddy which tends to grow; as part of this growth we have to reckon with its starting other eddies from its own resembling its own. This propensity it is which furnishes opportunity under the factors of evolution for a continual production of modified patterns of eddy. The patterns evolve some of them an increasing complexity. It is as though they progressed towards something. But philosophy reflects that the motion of the eddy is in all cases drawn from the stream, and the stream is destined, so the second law of thermodynamics says, irrevocably to cease. The head driving it will, in accordance with an ascertained law of dynamics, run down. A state of static equilibrium will then replace the stream. The eddies in it which we call living must then cease. And yet they will have been evolved. Their purpose then was temporary? It would seem so.'

We will return in a later chapter to an extension of this model of organisms. Here we may simply note that equilibria are often rather simply described states, at least from a macroscopic point of view (this is why systems in equilibrium are so popular with physical chemists), whereas open systems, like whirlpools or rabbits, are typically very complicated for so long as they last. Steady states can be very elaborate: they owe their persistence, not to being in a state of balance in an effectively unchanging world, but to being in a compensating state of inbalance within an environment which is itself in a maintained state of inbalance. The open system may owe its continued existence to more subtle factors than being a good compromise between energy and entropy effects: it may survive through being able to maintain a particular pattern of energy conversion—or perhaps because it can run quickly or hang on to a rock. Organisms and whirlpools have more things to do than just move towards a single 'ideal' (equilibrium) state.

We will start by considering crystals as models of the most elementary kind of material self-organisation: that which can result from a movement to a local equilibrium. In any case, such considerations are still important when discussing living systems even if they are not the whole story.

FURTHER READING

FAST, J. D. 1962. *Entropy*. Cleaver-Hume Press, London. For an elementary discussion of open systems, see OPARIN, A. I. 1957. *The Origin of Life on the Earth*. Oliver and Boyd, Edinburgh. p. 323. For a more advanced discussion, see PRIGOGINE, I., 1955. *Introduction to Thermodynamics of Irreversible Processes*. C. C. Thomas, Springfield, Ill.

The September 1970 issue of *Scientific American* is devoted to discussions of the cyclic movements of matter and energy in the biosphere.

PART ONE—FROM CRYSTALS TO ORGANISMS

2

Crystals

The characteristic regular arrangement of water molecules in ice below 0° is, we suppose, the best possible compromise between energy and entropy factors. Just above this temperature thermal motion is too violent for the ice arrangement to be maintained as an infinite array. Just below 0° the tendency towards an ordered state, in which the interatomic forces are more completely satisfied, becomes dominant. This is a broad thermodynamic explanation of freezing, in terms of a 'best possible' arrangement. But there is still a problem. How is this arrangement actually achieved? We must attempt also a *kinetic* explanation.

SUPERCOOLING AND SUPERSATURATION

If we cool water we may find that ice does not form at 0°C but only at several degrees below this temperature, and perhaps even then after a time lag. This is known as *supercooling*. Delays are also common when crystals are expected from solution: for example, liquid honey is a highly *supersaturated* solution of sugars which may take months to crystallise. These phenomena should not really be very surprising: the freezing of a teaspoonful of water requires the precise orientation of about 10^{22} molecules. The surprise is really that sizable crystals ever form in a reasonable time.

Some substances are exceedingly difficult to crystallise. Glycerine is a good example. Its freezing point is 18°C, yet it had been known as a liquid for centuries before it was discovered that it could be a solid. When glycerine does crystallise, however, it does so reasonably quickly. It happened on one occasion in a barrel being sent from Vienna to London (Oparin, 1957, p. 98). Gibson and Giauque (1923) found that after they had acquired some glycerine crystals, liquid glycerine samples in their laboratory would often crystallise apparently spontaneously. The atmosphere in their laboratory contained sufficient minute 'seeds' to do the trick. This is quite a common experience in organic chemistry: a new substance may at first be difficult to crystallise and then be very easy.

There seem, then, to be two parts to crystal formation. First, there is an invisible (slow) process of starting—in the appearance of the first suitably organised minute crystals (one may short-cut this process by 'seeding'). Then there is an eventually visible and commonly much faster process of growing.

Consider the boundary between a crystal and its liquid (say, between ice and water). Thermal agitation of molecules at the surface of the crystal will result in some of the momentarily more energetic of them flying off into the liquid. At the same time, collisions of suitably orientated molecules from the liquid with the surface will result, sometimes, in the addition of new molecules to the crystal. Whether the crystal as a whole will grow or dissolve will depend on which of these processes is occurring faster overall. At higher temperatures, the disintegration process will be dominant; at low enough temperatures the growth process will 'win' over the disruptive effect of thermal motion. The melting point of a solid corresponds to the temperature at which the two processes are in exact balance. For a given pressure this temperature is unique.

If ice is heated to just above 0°C, it will start to melt almost immediately. This is because the process of melting does not depend critically on the pre-existence of liquid. The process of crystal growth, on the other hand—the progressive accretion of molecules onto a crystal surface—requires the pre-existence of such a surface. Hence the need for a 'seed' or *nucleus* of some kind for crystallisation to get under way.

NUCLEATION

Surfaces generally may act as nuclei. For example, the sides of the containing vessel, or dust particles. But 'spontaneous' nucleation is also possible through the chance formation in the liquid of a suitable aggregate of molecules acting as a base for the growth processes. Spontaneous nucleation is likely only under fairly highly supercooled or supersaturated conditions. This is because very small crystals are much less stable than big ones. If, say, only 12 naphthalene molecules are arranged into a tiny 'crystal', almost all of the molecules will be at edges, incompletely surrounded—and hence incompletely held—by other molecules. A crystal is a co-operative enterprise whose overall stability depends on the mutual cohesion of a fair number of molecules. A result of this reduced stability of very small crystals is that they will melt at lower temperatures than the bulk crystalline material. So the temperature at which the crystallisation of a liquid will get under way is inevitably limited by the melting points of the small crystals which must appear before larger and more stable ones can be formed through their growth.

At any temperature, small aggregates of molecules in a liquid will be forming continually through chance encounters between the molecules. Above the melting point these will always re-dissolve. As the temperature is reduced, the average size of the aggregates will increase until at some rather indefinite stage an aggregate is formed which is large enough, and well enough organised, to grow spontaneously rather than re-dissolve: that is, roughly speaking, an aggregate is formed whose melting point is above the ambient temperature. This nucleus grows larger, its melting point rises as it thus becomes more stable, and the chances of its now re-dissolving become increasingly remote. The whole liquid then crystallises. The temperature at which this starts corresponds, not to the melting point of the bulk crystalline material, but more closely to the melting point of the critical 'seed' nucleus. But such a crystal nucleus is always to some extent a product of chance. Although as an aggregate grows bigger, under supercooled or super-saturated conditions, its chances of surviving improve, the odds are nevertheless against it until it reaches the 'critical size'. If the temperature is just below the melting point of the material, this size may require that hundreds of molecules happen to associate against the odds. It may require a freak of an event-sequence.

How long we are likely to have to wait for a crystallisation process to get under way spontaneously depends inversely on the volume of the liquid from which it is hoped the crystals will appear (Hinshelwood, 1951, p. 83). We would expect this for a process depending on a rare molecular event-sequence, since the larger the volume the greater the chances that a suitable event-sequence will occur somewhere within the volume, within a given period.

CRYSTAL GROWTH

Now let us think more carefully about the processes involved when new molecules add to a pre-existing crystal. Consider an idealised example, a collection of cubic 'molecules' packed into an incomplete 'crystal' as in Fig. 2. We suppose that this crystal is in contact with its slightly supersaturated vapour. The cohesion of a molecule on the crystal surface will depend on its situation. For example, molecule A is held only by one face, while B in the 'step' has two contact faces, and C in the 'kink' has three. A molecule landing on the crystal surface is most likely to be in a situation like A. It may well just skate about on the surface momentarily and then fly off again. However, if it encounters a 'step', then it will tend to be held there more firmly, with a sideways sliding motion. Again

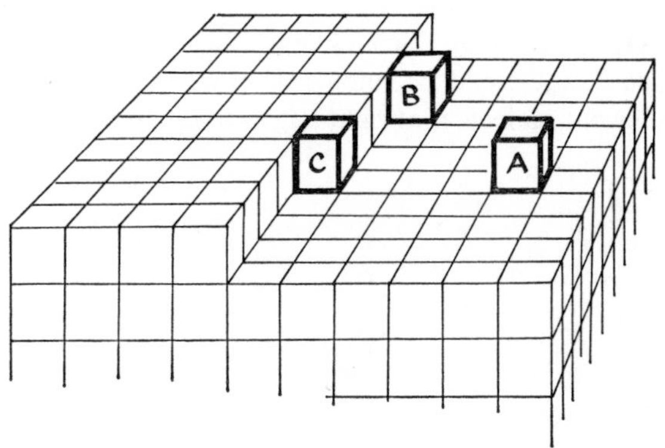

Fig. 2. The growth of a crystal face is assisted by the existence of a step (see text).

it might release itself, or alternatively it might encounter a 'kink'. Under supersaturated conditions the energy–entropy balance will favour its staying there. The crystal will thus grow through a progressive (net) addition of molecules at 'kinks'. A special problem arises, however, when a layer of molecules is completed, since this eliminates any steps. We need a happy accident, i.e. enough molecules in the A situation to happen to come together to form a stable island, thus creating new steps. Under only slightly supersaturated conditions, however, such accidents would be very rare—so rare that crystals would only grow imperceptibly slowly. This problem of starting a new layer was pointed out by Gibbs in 1878. Only within the last 25 years has it become clear how crystals can grow at a reasonable rate even in the presence of pre-existing 'seeds'.

CRYSTAL DEFECTS

According to the theory proposed by Frank (1949), rapid crystal growth depends on the existence of particular kinds of stacking errors—physical defects—within the growing crystal.

Broadly speaking, three kinds of physical defects can be found in three-dimensional crystals; these are called point, line and plane defects respectively. A localised imperfection such as the absence of a molecule—a vacancy—is a point defect (Fig. 3*a*). As the name implies, a line defect has an indefinite extension in one dimension. One might think first of a regular row of vacancies extending through a crystal. Such a situation would be rather unstable, however, since vacancies tend to wander about in a crystal in an arbitrary gas-like manner. Such a line would soon diffuse away. This kind of line defect could be stabilised, however, through displacement of adjacent molecules at right angles to the line to form an *edge*

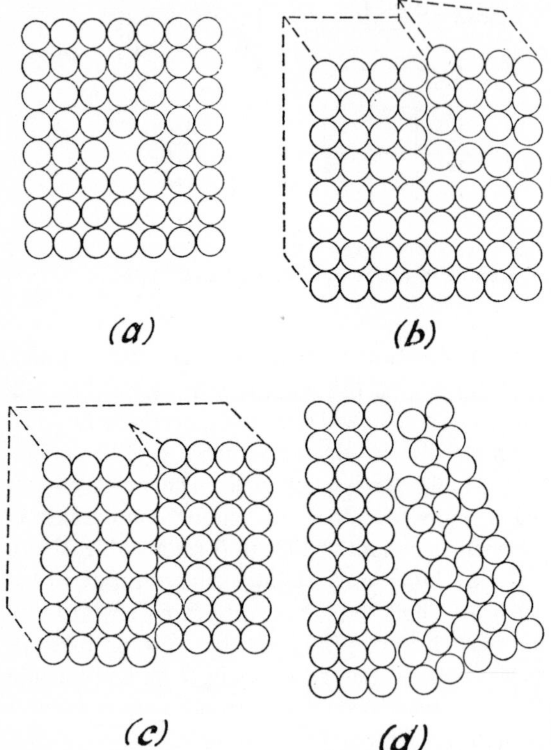

Fig. 3. Four kinds of physical imperfection in crystals.

dislocation (Fig. 3*b*). A *screw dislocation* can result from a displacement of lattice units parallel to a line in a crystal (Fig. 3*c*). A plane defect consists of an interface between two differently orientated crystals (Fig. 3*d*).

DISLOCATION THEORY OF CRYSTAL GROWTH

The surface of a crystal containing dislocations will not be entirely flat but will have various kinds of 'steps' which may operate to

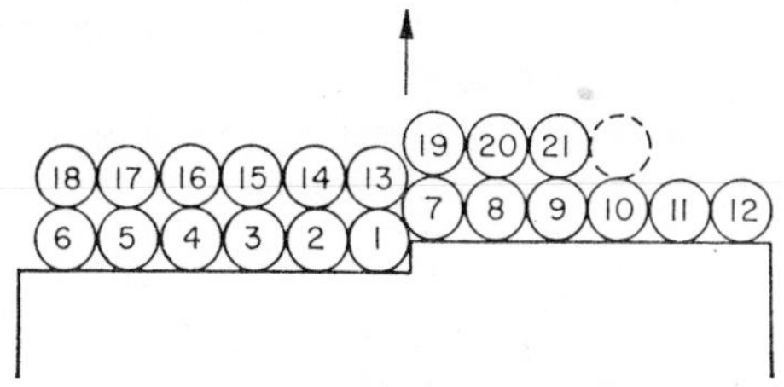

Fig. 4. The replication of a dislocation may provide a mechanism for rapid crystal growth.

assist crystal growth in the way which has already been described. Unlike the earlier mechanism, however, the steps are not so easily eliminated. Figure 4 illustrates one rather simple way in which a step may tend to be reproduced as the crystal grows.

The screw dislocation can provide a particularly effective mechanism for growth. Here the problem of starting a new layer is simply avoided—by there being no 'new layers'. The whole crystal stack is *one* layer formed into a spiral ramp—rather like some vertical car parks (Fig. 5). Again, the effectiveness of these

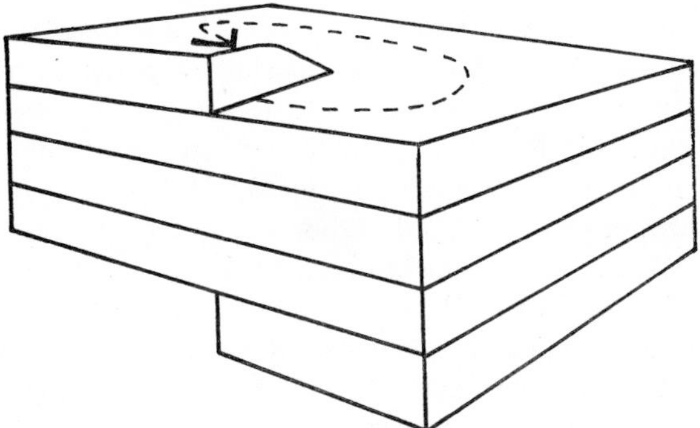

Fig. 5. The step created by a screw dislocation is not eliminated as the crystal grows.

dislocations in promoting crystal growth depends on their continual replication through the processes of growth. Later we will consider this in connection with biological reproduction.

CHEMICAL IMPERFECTIONS

So far we have been considering 'mistakes' in the *packing* of units in crystals. But often there can be 'mistakes' in the units themselves. You might say that a pure crystal of copper is chemically perfect. But copper crystals can be 'contaminated' with up to 30% of zinc without undue disturbance to the crystal lattice: the zinc atoms simply stand in for copper in a seemingly arbitrary way. Brass is only one of many examples of such *substitutional alloys* (see Fig. 6*a*).

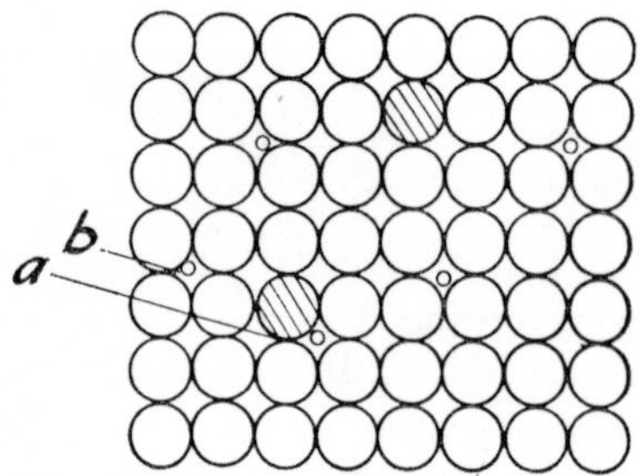

Fig. 6. Two kinds of chemical imperfection in crystals.

In some cases the substitutions in such alloys are *ordered.* For example, in a brass consisting of a 50:50 mixture of copper and zinc, at a temperature below about 460°C, the two kinds of atom alternate regularly in the lattice, each copper atom being surrounded by zinc atoms and vice versa. Above about 460°C the regularity disappears and the atoms distribute themselves randomly.

Such phenomena are not confined to alloys. For example, crystals of potassium dihydrogen phosphate contain hydrogen bonds in which there are two possible positions for the hydrogen atoms. At temperatures below −145°C all of these are in one of the alternative positions. Above this temperature they are probably distributed randomly between them (Bacon and Pease, 1955). A similar disordering of hydrogen atoms on hydrogen bonds occurs in crystals of 1-1′-methylene-di-2-hydroxyacridine giving rise to a mixture of isomers in the solid. These crystals become increasingly ordered at lower temperatures (Cairns-Smith, 1961).

Substitutions, notably aluminium for silicon, are a characteristic feature of silicates. Such substitutions may be more or less ordered and are frequently very stable. We will come to discuss this further in Chapter 7.

Another kind of chemical imperfection is illustrated by steel, which is an *interstitial* alloy. The small carbon atoms do not replace iron atoms in the lattice but fit into spaces between them (Fig. 6*b*).

FURTHER READING

The following books include accounts of crystal growth mechanisms.

GILMAN, J. J. (Ed.) 1963. *The art and science of growing crystals.* Wiley, New York.

VERMA, A. R. and KRISHNA, P. 1966. *Polymorphism and polytypism in crystals.* Wiley, New York.

See also

THOMAS, J. M. 1970. The chemistry of deformed and imperfect crystals. *Endeavour* **24,** No. 108, p. 149.

3

Subcrystals

In crystals of organic molecules, for example naphthalene (Fig. 7), we can distinguish between two levels of structure. First, there is the structure of the molecule: this is distinct and persistent through the 'faithfulness' of covalent bonds. Secondly, there is the structure of the crystal. This is also distinct and persistent under suitable circumstances, but here the precision of arrangement arises from the operation of a number of delicately balanced attractive and repulsive tendencies: the complex van der Waals forces tend to pack the molecules together as neatly as possible; the space occupied by the electron clouds of the molecules nevertheless prevent too close an approach; thermal vibration all the time insists on some elbow room. Collectively we may call these 'packing factors'. Because the attractive forces are weak, a molecular crystal structure only begins to hang together when a fair number of molecules are assembled.

Even for simple molecules like naphthalene it is difficult to be sure how they will pack. For a more complex one like tryptophan:

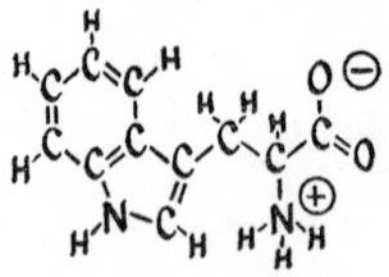

the problem is even greater. Five different chemists could well have five different theories as to how some newly discovered molecule should pack itself in the crystal. But five different batches of the molecules themselves would (probably) agree quite quickly as to precisely the best answer.* In any case they would array themselves with a virtually endless regimentation in all directions like a vast three-dimensional wallpaper.

---

* According to Kitaigorodskii (1961, 1970) close-packing is by far the most important factor in determining a crystal structure, and prediction may be possible on this basis. But in any case the technique of X-ray crystallography allows us to discover the answers that molecules find for themselves (see Bragg, 1968). For crystal structures of amino acids like tryptophan, see Marsh and Donohue, 1967.

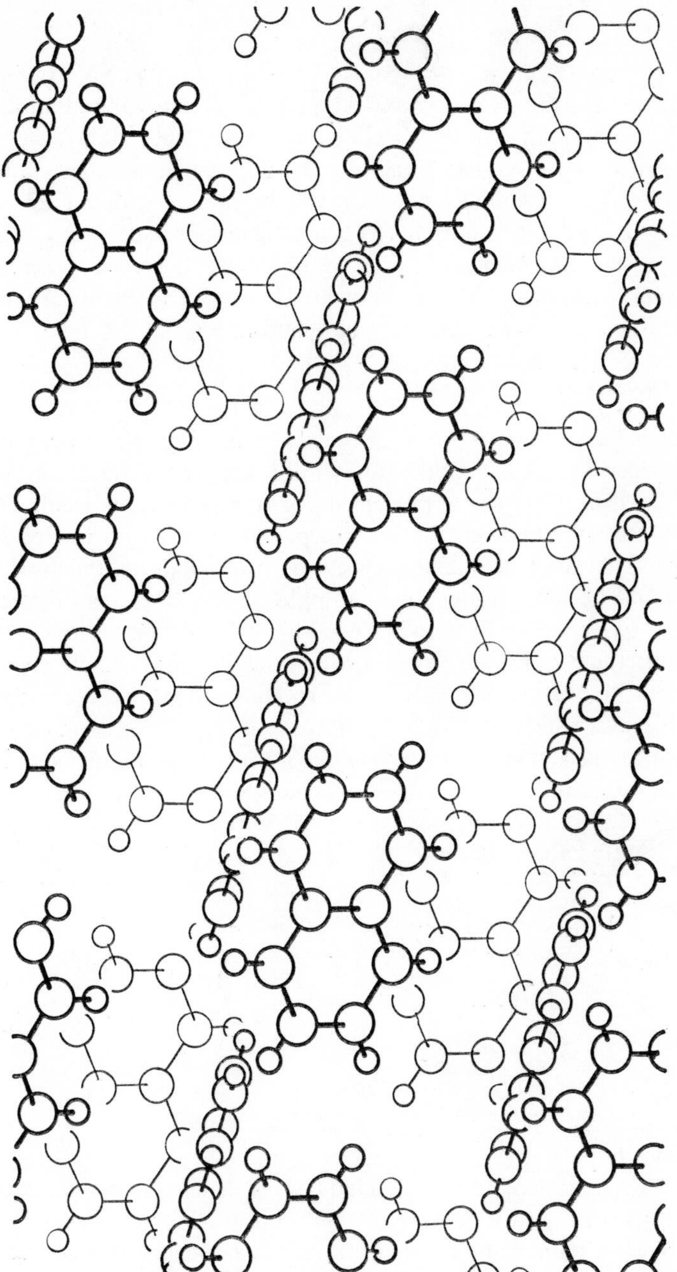

Fig. 7. Arrangement of molecules in a crystal of naphthalene.

The crystal, then, is a frequent consequence of the operation of packing factors. But it is not the only kind of organised structure which can so arise. Here we will use the term 'subcrystal' to describe systems whose organisation depends, at least in part, on packing factors, but which fall short of the usual idea of a crystal in not being infinite in all dimensions, in not being regular, or in deviating in both of these respects. Bernal (1965, 1966) has considered a 'generalised crystallography' which would include many such systems. They are of pre-eminent importance in molecular biology.

CONFORMATIONS

To make this connection with molecular biology, we must start by considering a rather subtle distinction which organic chemists make when discussing the arrangement of atoms in a molecule. This distinction is between the relative positions of atoms immediately joined to each other and the relative positions of atoms that are more distant. Let us consider some examples. For a very small molecule, such as water or methane, the covalent bonds fix within narrow limits the relative positions of the atoms in space. Single covalent bonds, however, normally allow rotation about their axes; thus molecules which contain a chain of more than three atoms may be able to exist in various *conformations*. (See, for example, Lambert, 1970). In ethane, for example, the relative orientation of the triplets of hydrogen atoms attached to the two carbon atoms can be altered by rotation about the C—C bond:

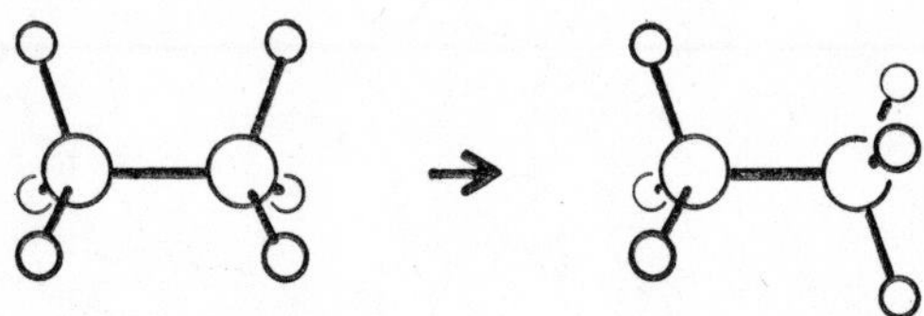

Molecules of hexane can twist in very many different ways (see opposite page).

Although the individual bonds resist bending and stretching, because of the easy rotation the molecule as a whole is quite floppy.

We see, then, that a covalent structure is not necessarily a complete account of the spatial relations between the constituent atoms. You might say that your tailor fixed the 'covalent structure' of your suit;

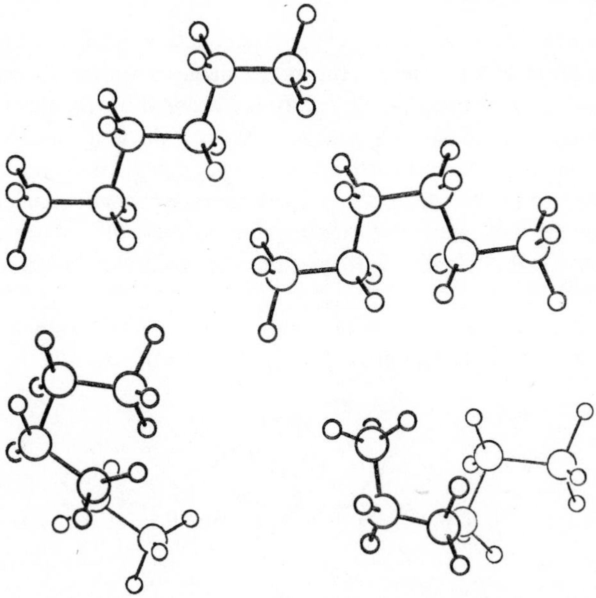

but its conformation at any moment depends on how your limbs happen to be disposed. Molecules may contain special covalent features which fix or limit their conformation (for example rings as in naphthalene). Non-covalent forces can also fix or limit conformations. This is particularly true for very large molecules. The conformation of a polymer may be strongly affected by hydrogen bonding, ionic bonds, or van der Waals forces operating between different parts of the molecule. Such forces operating between a dissolved molecule and its solvent may also be important. For example, in solutions of polymers consisting of very long floppy chains, the molecules may tend to bunch up or assume more extended conformations according to the solvent. In a good solvent—where the attractive forces between polymer and solvent are relatively strong—extended arrangements will tend to be preferred, since this allows the maximum possible interaction between polymer and solvent. In poor solvents, on the other hand, attractive forces operating between different parts of the same molecule may predominate—leading to bunched conformations.

POLYMER CRYSTALS

You could think of a chain lying in a heap as a model of a bunched conformation of a polymer. This model suggests another possibility. If the links of a chain are identical, or if there is some regular pattern of repetition of links, then it is possible to fold it up neatly. You could pleat it back and forth for example. Now, the same factors that give rise to the neat packing of small independent molecules in ordinary crystals may give rise to neat folding of polymers; not simpy to a 'bunched' conformation but to a 'crystalline' conformation.

Consider polythene. This molecule is made by joining together hundreds of thousands of ethylene molecules into a chain:

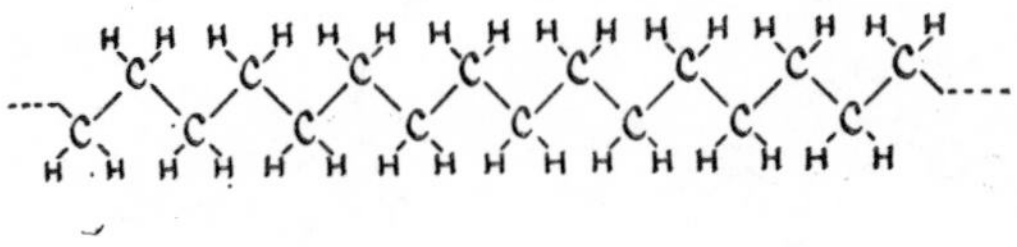

When molten polythene cools, it becomes increasingly opaque through myriads of orderly 'crystalline' regions forming throughout the material. In these regions limited stretches of chains have packed together in register, as on the left-hand side of the diagram:

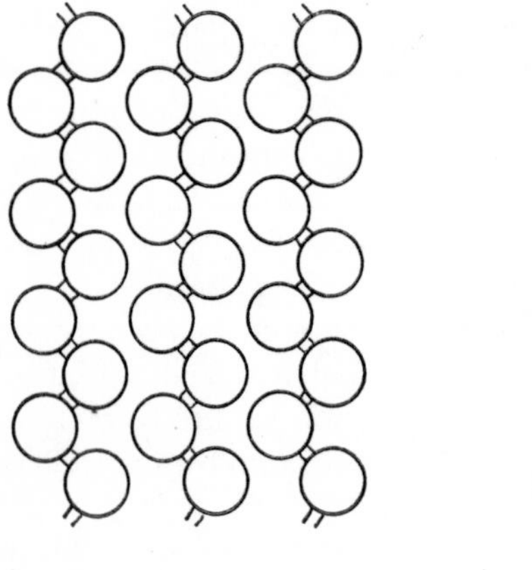

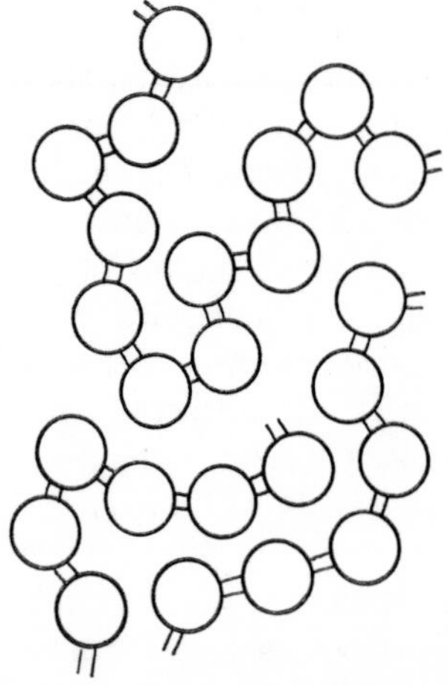

If polythene is allowed to crystallise from dilute solutions, then highly geometrical 'single crystals' may be formed (Keller, 1957; Fischer, 1957; Till, 1957). These consist of very thin platelets—about a millionth of a centimetre thick (see Plate I*a*, *b*). In these platelets the polymer molecules are stacked with the chains upright:

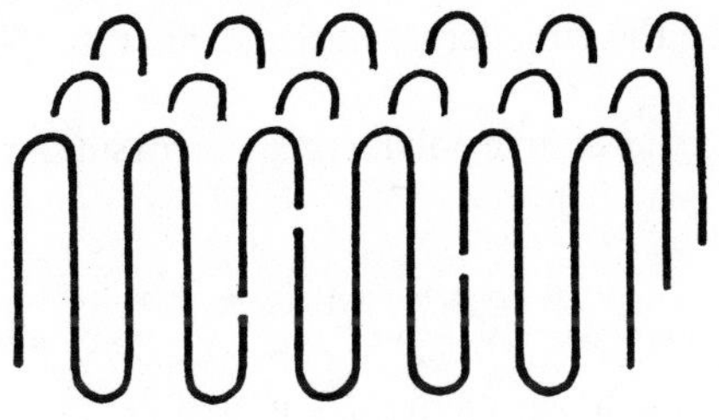

(Since the average length of the chains is many times the thickness of the platelets, the chains must fold back and forth in some such way as indicated above.) The platelets, then, consist of parallel ribbons, each ribbon being formed by a pleated chain.

There is, however, an additional subtlety. It seems (Keller, 1962) that both the pleating of the chains to form ribbons, and the stacking of the ribbons, are slightly oblique (see Fig. 8). Both of these oblique

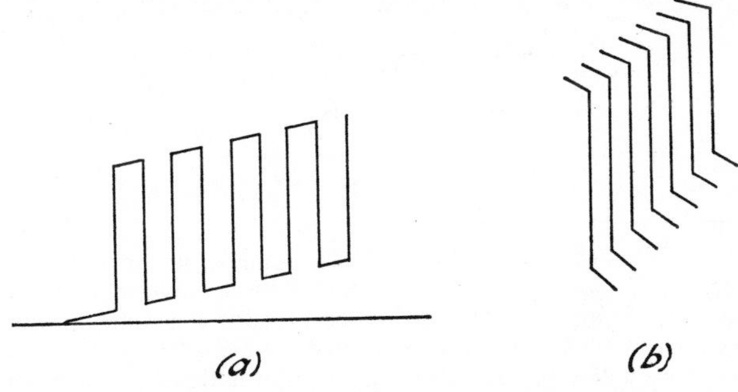

Fig. 8. (*a*) The polythene chain pleats to give a slightly oblique ribbon. (*b*) The edge view of a ribbon stack shows a second kind of climb angle.

effects are probably due to a necessary assymetry in the conformation of the fold regions (Frank, in Keller, 1962), i.e. the folds are not only lop-sided when viewed as in Fig. 8*a*, but also when viewed from the side as in (*b*) where they lean one way or the other. A result of these packing features is that, instead of being flat, the platelets may be formed into three-dimensional microscopic objects, typically hollow pyramids (Plate II).

Notice the hierarchical organisation of these structures: the line pleats to form a ribbon; the ribbon folds round to form a platelet; such platelets may then stack to give a composite 'crystal'. Bernal (1959) has stressed that a hierarchical building up of forms is characteristic of the submicroscopic and microscopic organisation of living cells.*

A number of other simple polymers can form similar single crystals, e.g. nylon:

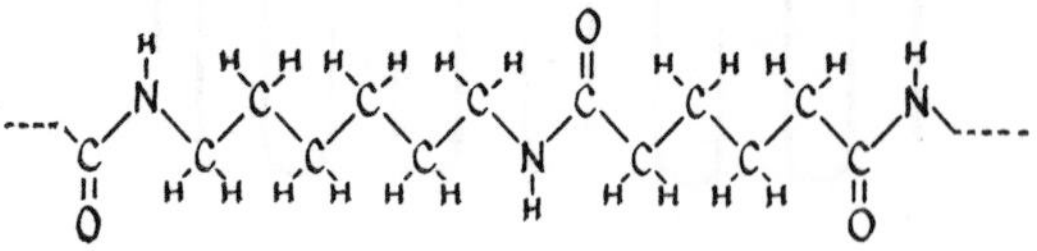

POLYAMINO ACIDS

The most uniform way of folding a string is not in a pleated but in a *helical* conformation. Only in this way is every part of the string in an identical situation—at least if you ignore the ends. Not surprisingly, regular polymers often take up such a conformation. Particularly important examples are found among the poly-α-amino acids which have the following general structure:

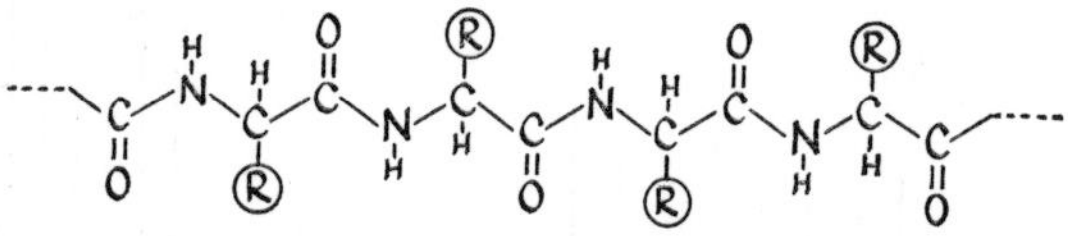

* Later in this chapter we will consider an example, in simple viruses, where different levels of folding and association of protein can give rise to a geometrical box. But there is an important difference between a pyramid formed from polythene and a virus box formed from protein. In current jargon we might say that polythene, unlike protein, contains no information. Polythene molecules happen to fold into pyramids for purely physico-chemical reasons: if virus protein associates to form a box, this is mainly for a biological reason—because the box is useful to the virus. No doubt we will learn to control the formation of the microscopic objects that can arise from the self-folding of polymers in the way that natural selection has come to control the formation of virus boxes. We will develop ways of synthesising essentially simple molecules like polythene or nylon which contain fold-inducing features—'cues'—such as bulky groups inserted at exactly predetermined positions (see Cairns-Smith and Pettigrew, 1969). Thus we might be able to make not only things like hollow pyramids, but boxes, tubes or other colloidal objects that might be useful to us. In particular such 'colloidal engineering' could form a basis for the design of specific catalysts.

where ® represents a group of atoms. ® is different for different polyamino acids. A long chain of this sort can twist and fold in a vast number of different ways. There are, however, factors which operate in favour of a particular conformation—the so-called α-helix. The special stability of this conformation arises largely from N—H · · · O hydrogen bonds between adjacent turns of the helix. If we simplify our representation of a polyamino acid to show just the hydrogen-bond-forming groups:

```
     O   H         O   H         O   H        O
     ‖   |         ‖   |         ‖   |        ‖
-N-C —N-C —N-C —N-C —N-C —N-C —N-C-
 |       ‖   |       ‖   |       ‖   |
 H       O  .H       O   H       O   H
 1       2   3       4   5       6   7
```

then the helix arises through hydrogen bonding of N_1 with O_4, N_2 with O_5, N_3 with O_6, N_4 with O_7, and so on:

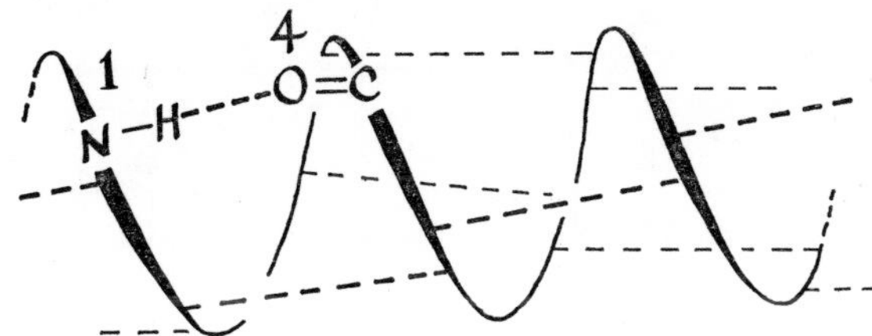

If we now represent the above by a simple cylinder, then the ® groups stick out sideways like a helical array of pegs on a post:

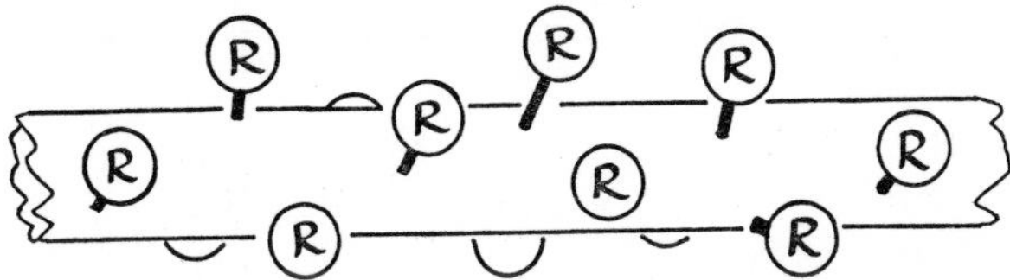

The α-helix is a one-dimensional subcrystalline arrangement: it is called a *secondary structure.* It has essentially the same *raison d'être* as a crystal structure—it can, for example be 'dissolved' by suitable solvents (see Doty *et al.*, 1956, 1957; Zimm and Bragg, 1959) which do not affect the primary (covalent) structure.

THE PRIMARY AND SECONDARY STRUCTURES OF DNA

The primary structure of deoxyribonucleic acid (DNA) consists of a chain to which are attached four different kinds of side groups—known as 'bases':

Generally there is no evident regularity in the arrangement of these side groups along the chains which may be millions of units long. A representative fragment of a DNA primary structure is shown in detail in Fig. 9.

In 1953, Watson and Crick suggested a secondary structure for DNA. They proposed that the molecules exist as double chains held together, at least partly, through hydrogen bonding between the side-group bases. The molecule thus has a sort of ladder structure:

Now, such pairing of the side-group bases requires (i) that there is a suitable plug–socket relationship between the hydrogen-bond-forming groups of the paired bases, and (ii) that a small base pairs with a big one, and vice versa. This specifies completely the possible pairs. Thymine must go with adenine, and cytosine with guanine:

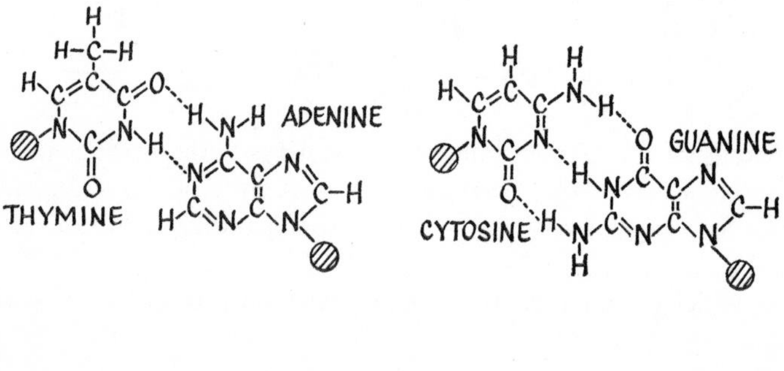

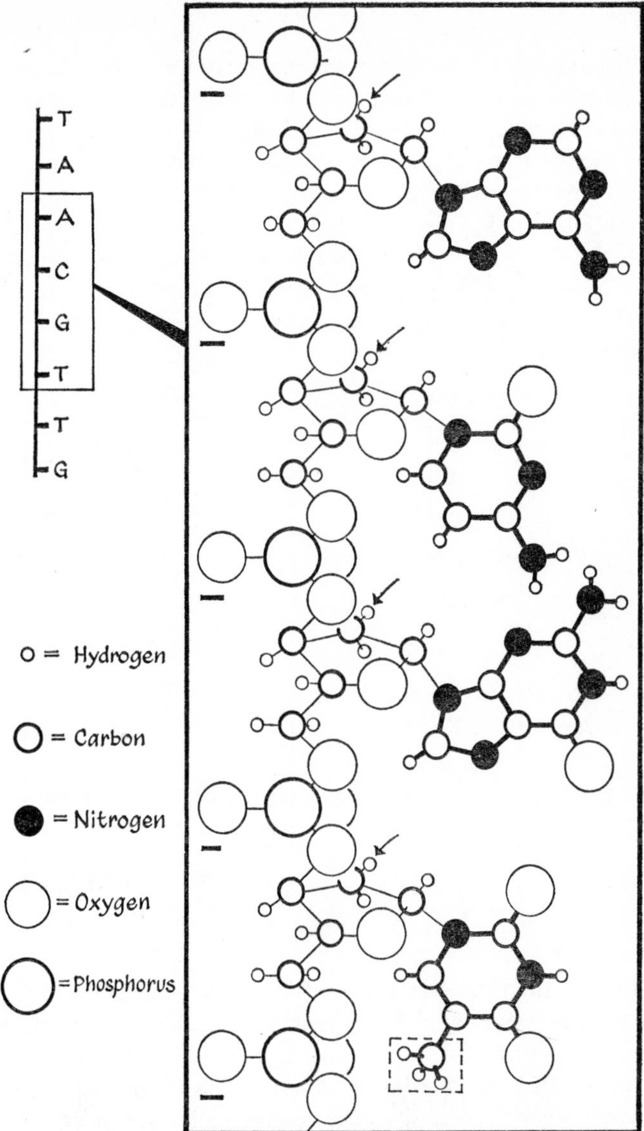

Fig. 9. DNA primary structure. The four bases, adenine, cytosine, guanine and thymine, are flat (naphthalene-like) units and are shown with thick bonds to distinguish them from the backbone of sugar and phosphate units which hold the bases in a particular sequence.

RNA primary structure is very similar: the hydrogen atoms indicated by the arrows are replaced with O—H groupings, and the base thymine is replaced by uracil, which has a hydrogen atom in place of the (boxed) methyl grouping.

It follows that if this secondary structure is to form, then the primary structures of the two chains must be strictly complementary. The sequence in which the bases are arranged in either of the chains must specify completely the base sequence in the other chain.

The complementary character of the DNA structure suggested to Watson and Crick a general kind of mechanism through which a specific DNA isomer could duplicate—through the chains separating, individual A T G C units pairing with bases on the separate chains, and these units then zipping together to give two pairs of chains identical with the original pair:

```
├A··T┤         ├A            T┤
├T··A┤         ├T            A┤
├T··A┤   →     ├T            A┤
├G··C┤         ├G            C┤
├C··G┤         ├C            G┤
                             ↓
├A··T┤  ├A··T┤       ├A··T–      –A··T┤
├T··A┤  ├T··A┤       ├T··A–      –T··A┤
├T··A┤  ├T··A┤   ←   ├T··A–      –T··A┤
├G··C┤  ├G··C┤       ├G··C–      –G··C┤
├C··G┤  ├C··G┤       ├C··G–      –C··G┤
```

Although the processes involved in DNA duplication are still a matter for active research (see Chedd, 1970), this mechanism seems to be at least formally correct. In any case, it is now clear that the ability of DNA to duplicate forms the basis for the ability of organisms to reproduce: it is the central molecular process of life. We will return to the question of the function of DNA in the next chapter; meanwhile we must give a final twist to our description of its secondary structure.

We can see more clearly why the bases lock together in such a specific way if we consider more carefully how the DNA strands, and the connecting base-pairs, are arranged in space. One can construct a rough kind of model, illustrating the general idea, out of wooden slats—representing the base *pairs*—and strings—representing the sugar–phosphate chains. These are arranged to form a ladder in which the slats are at right angles to the strings. By twisting the strings into a double helix—like an electric light flex—the slats will approach each other until finally they are in contact:

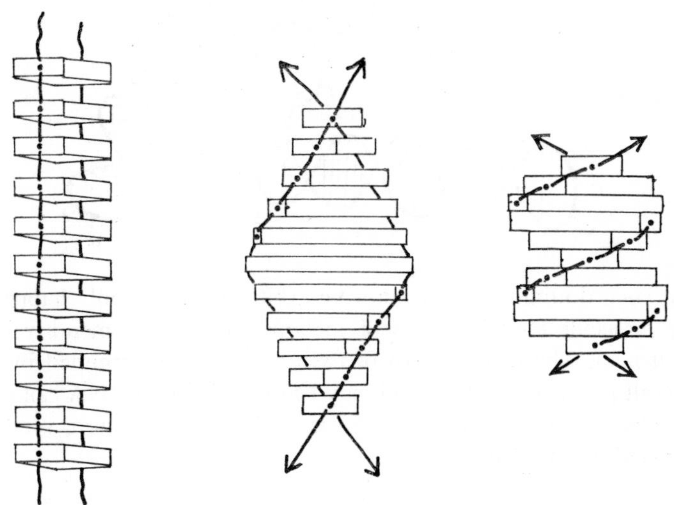

As in many ordinary crystals, the final packing arrangement is probably a compromise between various factors: for example, base-pair hydrogen bonding; van der Waals forces between the flat base-pairs; ionic and polar interactions tending to keep the sugar–phosphate chain on the outside in contact with water; and also purely geometrical factors restricting the space within the helix.

RNA

Ribonucleic acid (RNA) has a very similar primary structure to deoxyribonucleic acid. The points of difference are (i) there is an O—H grouping instead of an —H atom at one repeating position on the main chain, and (ii) the $—CH_3$ group in thymine is replaced by an —H atom. This simpler base is called uracil (see Fig. 9, p. 29).

Secondary structure in RNA is less definite than in DNA. The molecules appear to exist as single strands which may fold back on themselves—like the result of strongly twisting a piece of string and then letting go the ends:

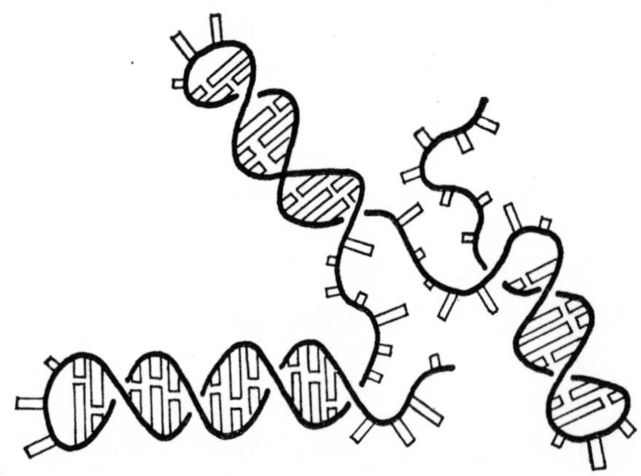

But such doubling back cannot be arbitrary. It can only happen if the molecule contains pairs of sections with complementary base sequences. The final pattern of twisting is thus determined by the overall primary sequence and may be quite definite: RNA can have a *tertiary structure*. As we shall see, RNA is thus a more versatile molecule than DNA.

PROTEINS

The most versatile kind of polymer in organisms—indeed, the most versatile substance that we know of—is protein. The quills on a porcupine are made of protein. So is silk—and wool, and hair, and muscle, and tendon, and gelatin, and the white of an egg. And so are the thousands of different specific catalysts which guide the complex nets of chemical reactions which we associate with 'living matter'. Not only are these *enzymes* proteins, but so are the antibodies which can accurately recognise and eliminate invading micro-organisms in higher animals. And there are many other examples of fundamental biological processes which depend critically and directly on proteins, e.g. for the eye to register the arrival of a particle of light, or for a nerve impulse to be transmitted, or for oxygen to be transported efficiently in the blood, or for a muscle to contract, it is necessary for proteins, differently made proteins, to interact specifically with other molecules or ions. Proteins really are

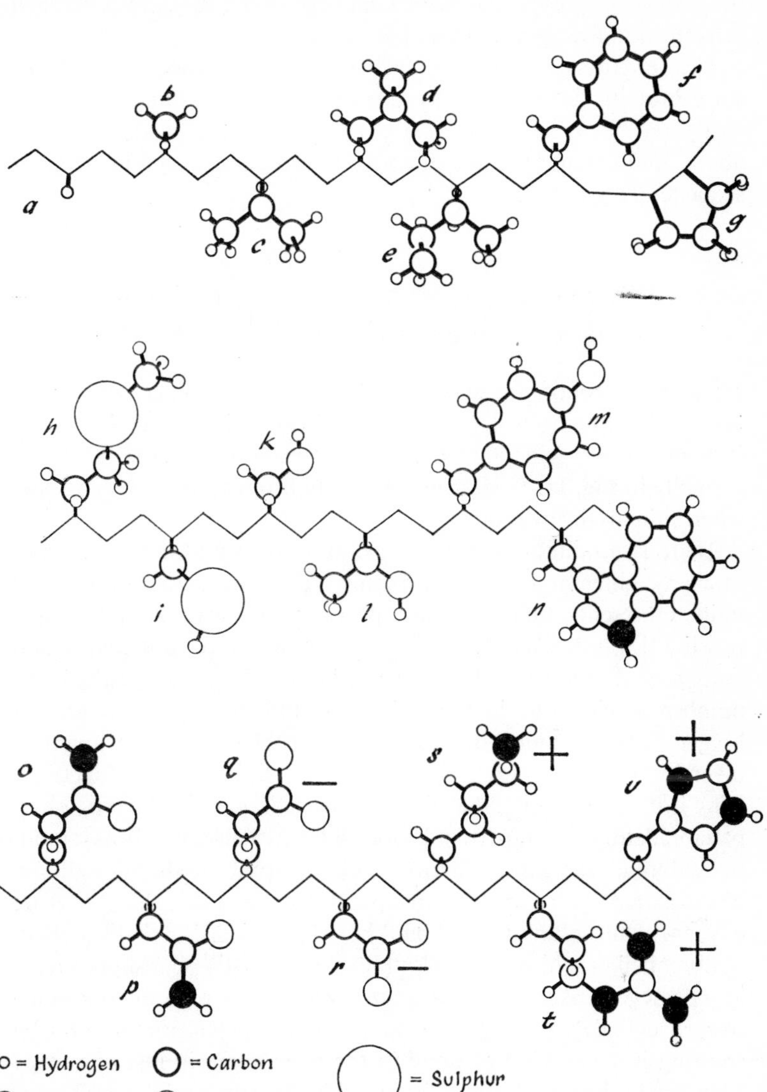

Fig. 10. The protein alphabet. NB. The amino acid side groups have been arranged in a manner that emphasises relationships between them, (page 37). The assignment of letters does *not* correspond to the recently proposed convention (in *European Journal of Biochemistry*, vol. 5, 1968, pp. 151–153).

the 'works' of present day terrestrial organisms: they are concerned directly or indirectly with every biological process.

The functional diversity of protein is all the more remarkable since proteins have what may seem to be a rather limited kind of structure. All proteins are 'mixed' poly-α-amino acids in which about 20 different kinds of side groups can be arranged in different sequences, e.g.:

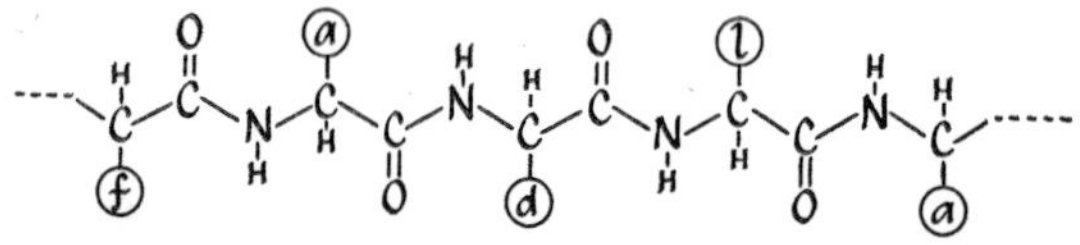

These side groups are shown in Fig. 10. On boiling with water and acid, proteins will eventually hydrolyse, i.e. break into constituent amino acids through adding water molecules as shown formally in Fig. 11. Similar processes are brought about by enzymes when animals digest protein.

There is no difficulty in principle in accounting for the existence of many different kinds of proteins. A chain of 150 amino acid units represents quite a small protein; but with 20 alternative possibilities for each link the total number of different chain sequences that are possible is 20^{150}, i.e. 10^{195}—far more than 'the number of electrons in the universe'—and most proteins are far longer than 150 units. But the central problem remains: how can diversity of sequence give rise to such a rich diversity of function? Why is it, for example, that the sequence cdkqaqnodcducmbscqbrcbauaoredetdfskugqldqsfrtfsudslqbqhsbkqrdssuaclcdlbdabedsssauuqbqdsgdbokublsusegesmdqfekqbeeucduktugapfabrboabhpsbdqdftsrebbsnsqdamoa (see Fig. 10) corresponds to a molecule that can store oxygen, while qlbbbsfqtqlkkpuhkbbkkkpmjpohhsktpdlsrtjsqcplfcuqkdbrcobcjkospcbjspalpojmokmklhkelrjtqklaksmgpbjmsllpbosueecbjqapgmcgcufrbkc, although quite incompetent as a one-molecule oxygen cylinder, is very good at breaking up RNA molecules by splitting just one kind of bond in the main chain in just one way? Then again, how is it that scfatjqdbbbhstuadrpmtamkdapmcjbbsfqkpfploblptplraklrmaedoepktnnjrpatlgaktpdjpegjkbddkkrelbkcpjbsseckrarahpbmcbntptjsalrcobnetajtd lacks any appetite for RNA but has a neat way of destroying bacteria by unstitching their overcoats? How can specific complex functions be carried out by such molecular cryptograms?

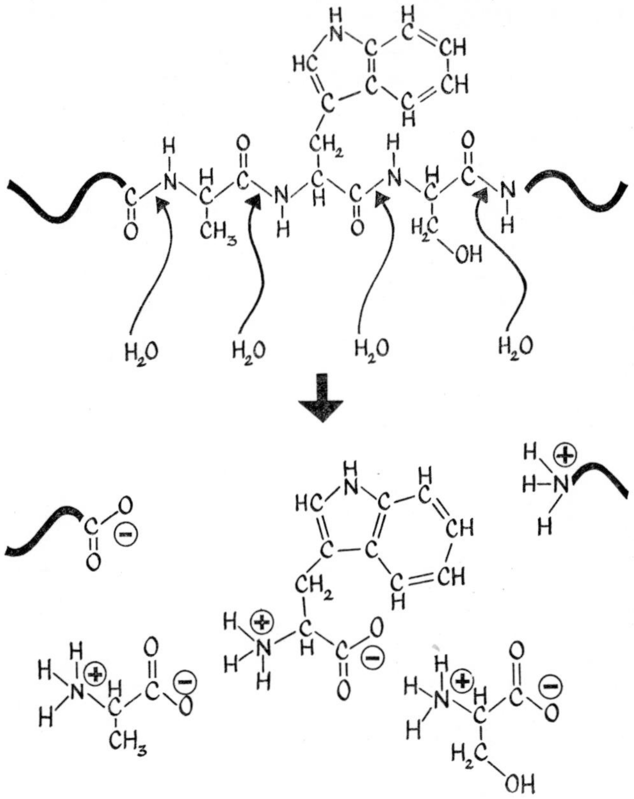

Fig. 11. Proteins can be broken into constituent amino acids by heating with water and acid. Here alanine, tryptophan and serine molecules are being formed through the hydrolysis of a small section of a protein chain.

The broad answer seems to be this: the 'cryptogram'—the primary structure—determines in detail the way in which the chain will collapse on itself, its tertiary structure. The 'cryptogram' may thus determine accurately the form of a piece of machinery about a millionth of a centimetre across.

IRREGULAR FINITE SUBCRYSTALS

Thus far, in previous sections, we have considered mainly regular subcrystalline polymer conformations. But perhaps such regularity is rather incidental—it arises from the repetitiveness of the primary structure. In ordinary crystals, too, one might suggest that the regularity is an incidental consequence of the rather banal way in which crystals are made—from vast numbers of only one or a few kinds of identical units (Fig. 12*a*).

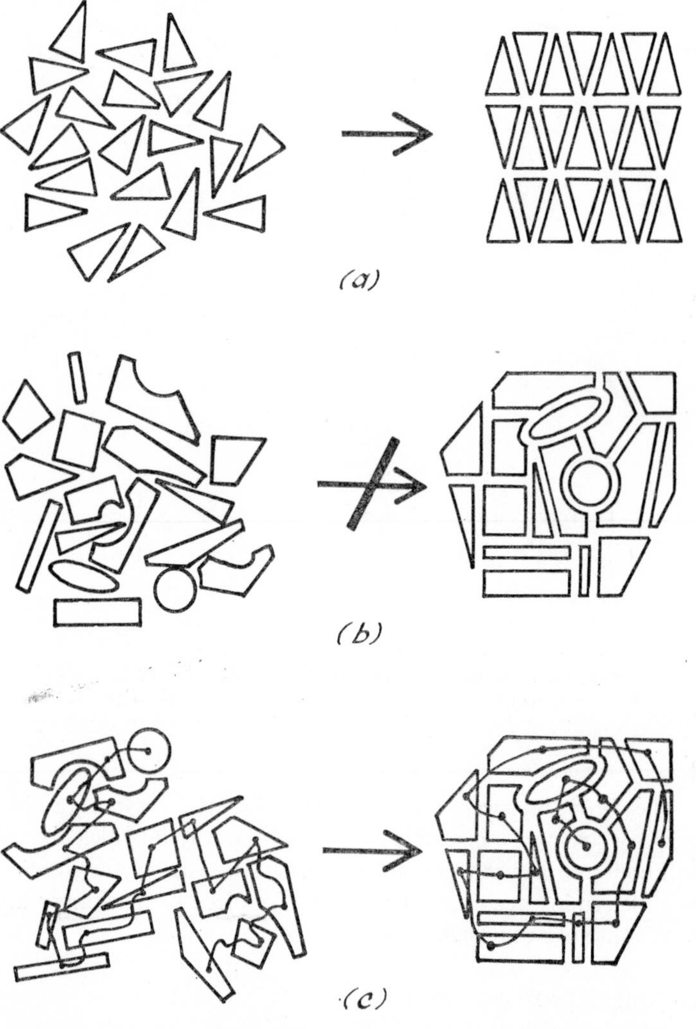

Fig. 12. (*a*) Under suitable circumstances a few dozen identical units may crystallise, but (*b*) with a complex mixture no preferred arrangement appears, unless (*c*) the units are joined together, in which case it becomes a possibility.

One might conceive of a 'crystal' constructed along more imaginative lines in which a few hundred different kinds of organic molecules of various shapes, sizes, and polarities are neatly packed, like a weekend suitcase, in some unique 'best possible' but quite irregular way (Fig. 12*b*). As far as we know, however, no such things exist. Such a very complicated arrangement might well represent the best mode of packing a mixture of molecules at some very low temperature. The trouble is, we would guess, that at normal temperatures the entropy factor would be unfavourable: there would be too few ways in which such a complicated structure could form, while there would remain too many ways in which it could come apart. So at normal temperatures such structures are not stable: they are, so to speak, being continually shaken apart faster than they can find ways of coming together again. On the other hand, at very low temperatures, at which such specifically complicated structures might be stable, they would form only imperceptibly slowly.

Yet, a particular kind of irregular finite subcrystal was discovered by organisms aeons ago—in their soluble protein molecules. The trick, it seems, is to tie the multifarious component parts together as a chain before attempting the 'subcrystallisation' (Fig. 12*c*). Figure 10 showed the rather bizarre collection of organic groupings which may be attached to the main chain of a protein. Some of these are, for example, hydrocarbon (*b–g*); some contain flat rings (*f*, *g*, *m*, *n*, *u*); two carry negative charges (*q*, *r*); three carry positive charges (*s*, *t*, *u*); there are three hydroxyl-containing groupings (*k*, *l*, *m*), and so on. According to the sequence of these groupings on the main chain, the protein molecule may fold into a complex three-dimensional 'globular' structure in which a large proportion of the groups fit into a very precise but irregular arrangement.

Myoglobin (see Kendrew, 1961) and haemoglobin (see Perutz, 1964) were the first proteins to be analysed in sufficient detail for their tertiary folding to become clear. Figure 13 shows the arrangement of the main chain in myoglobin. About 75% of this chain is in the form of straight α-helical sections. The chain changes direction in non-helical regions to give an overall effect of great complexity and irregularity. Haemoglobin consists of four chains, each with essentially the same tertiary structure as myoglobin. On the other hand, the enzyme lysozyme, which has also been analysed in detail (see Philips, 1966), contains much less α-helix; and it is known in

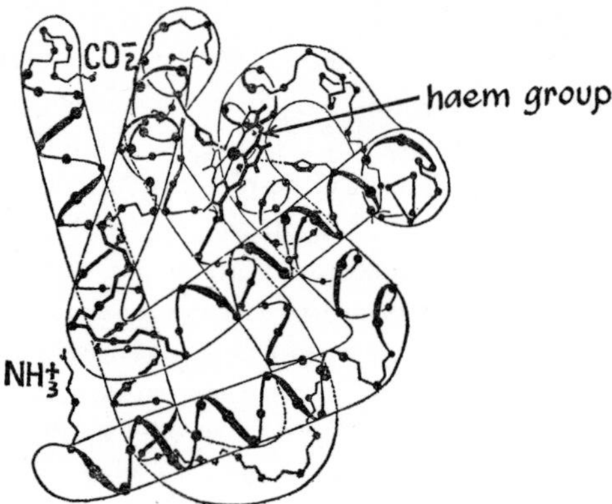

Fig. 13. The secondary and tertiary folding of myoglobin. The side groups (not shown) occupy most of the remaining space within the structure, leaving a slot into which the haeme molecule fits.

general that the helical content of proteins is highly variable. The most important factor in stabilising the overall conformation of the soluble proteins which have so far been studied appears to be the strong tendency for the hydrocarbon side groups to pack together neatly inside the molecule away from the surrounding water—and it is difficult to make any other generalisation (Kendrew and Watson, 1966).

Many soluble proteins have S—S bridges which can be formed by oxidation of pairs of SH groups situated at different points in the chain:

$$\text{—SH} \quad \text{HS—} \xrightarrow{\text{O}} \text{—S—S—} + H_2O$$

These would seem to be important in maintaining tertiary structure. But they are clearly not essential: neither myoglobin nor haemoglobin, for example, have any; and ribonuclease can refold spontaneously, to reform its correct bridging arrangement, after its S—S bonds have been broken (Anfinsen, 1963). It seems that the function of S—S bridges is to stabilise a tertiary structure which has already been established through the concerted effect of numerous, much weaker forces—that the precision of the tertiary structure arises not, as in a rigid molecule (like naphthalene), from covalent bonds, but (as in a crystal) from packing factors.

Let us return to our general problem: why is it possible for a 'mixed' polymer to form a complex subcrystalline organisation while such organisation seems to be impossible for a mixture of independent molecules? The answer is presumably because in a polymer, even in a quite floppy polymer, the number of ways in which the units can be arranged in space is enormously more limited. By joining together (say) 150 amino acids into a chain, you are not only holding the members of a particular collection in close proximity to each other but also, if you have joined the units up in a particular order, you are defining one set of next-neighbour relationships, between the units along the chain, out of a total of 10^{195} possible sets of such relationships. This is still, of course, far from a complete specification of the arrangement of the groups in space: the chain can be twisted in a vast number of different ways. But perhaps this number may be no longer *so* vast as to prevent the trial and error approach of random thermal motion being able *now* to hit on some particularly favourable packing arrangement, even if it is complicated, and being able to do so often enough for this arrangement to exist for a large proportion of the time at ordinary temperatures. (We might compare a random golfer who can sink a putt quite quickly because the ball has been tethered to the flag pole on a shortish string.)

There are two kinds of organisation involved, then, in arranging the atoms in a globular protein molecule. There is the physical self-organisation—the 'subcrystallisation'—which can lead to a conformation with a best possible energy–entropy balance. But really this arrangement has been more or less 'rigged'. It has been more or less preselected by the amino acid sequence. The origin of the primary sequential organisation in proteins is quite different from that of the subcrystalline organisation which may be superimposed on it. But we shall leave a further discussion of this until Chapter 6.

ON THE VIRTUES AND DIFFICULTIES OF MOLECULAR SOCKETS

Many proteins have an extraordinary ability to recognise other molecules: indeed, this is perhaps the main *raison d'être* of proteins in general. The ability seems to depend on protein being able to form 'sockets' which fit specifically only certain other molecular species. A protein may thus be able to select a particular molecule out of a mixture, since the molecule that fits the socket most exactly will be held most firmly.

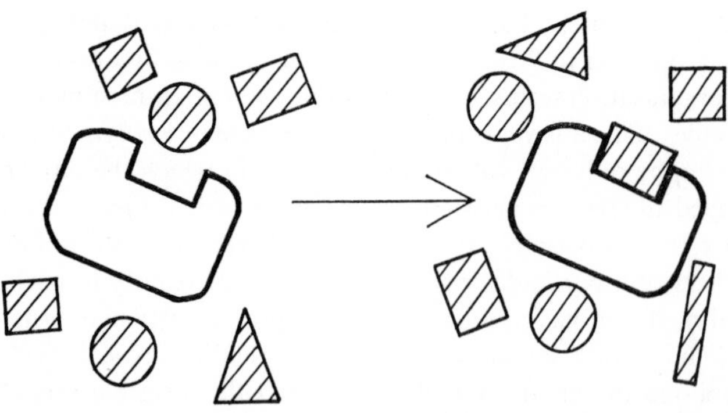

Myoglobin is an example. Myoblobin acts as a temporary oxygen store in the muscles. Its oxygen-binding power depends on the protein containing a slot which fits very exactly the shape of the flat, iron-containing molecule haeme (see Fig. 13). There are about 100 contacts between the haeme molecule and 13 different side groups within the protein (Kendrew and Watson, 1966).

Consider the problem of forming a structure that will unite selectively with some particular molecule which you are given, say, 2-methyl-naphthalene:

```
      H     H     H
      |     |     |
  H\ C=C\ C/C\ C/C-H
     |    ||   |   \
     |    ||   |    H
  H/ C\ C/C\ C=C\ H
         |     |
         H     H
```

You want a structure with a 2-methyl-naphthalene-shaped groove in it. With the aid of molecular models you might try to design some suitable molecule—perhaps a covalent framework consisting of a pair of naphthalene-like groups held apart at just the right distance (3·7 Å) and in some ingenious way able to accept only naphthalene molecules with a methyl group in the correct position. A job for a molecular locksmith. And perhaps not worth attempting—since there already exist structures which have a selective affinity for

2-methyl-naphthalene through having 2-methyl-naphthalene-shaped grooves in them: *crystals* of the stuff. These will have more or less shape-specific steps and kinks on their surfaces. (When a chemist recrystallises a substance to purify it, he is making use of this curious self-selective property of molecules.) So here is another angle of similarity between proteins and crystals: they represent the two major groups of specific molecular socket structures.

By having an irregular subcrystalline organisation, proteins are not so restricted as crystals in the kinds of socket which they can make: for example, protein sockets need not bear any resemblance to the units out of which their subcrystal is formed (cf. Fig. 14*a* and *b*). More cunningly, a protein may even be able to envelop another molecule (Fig. 14*c*). In either case we can think of the

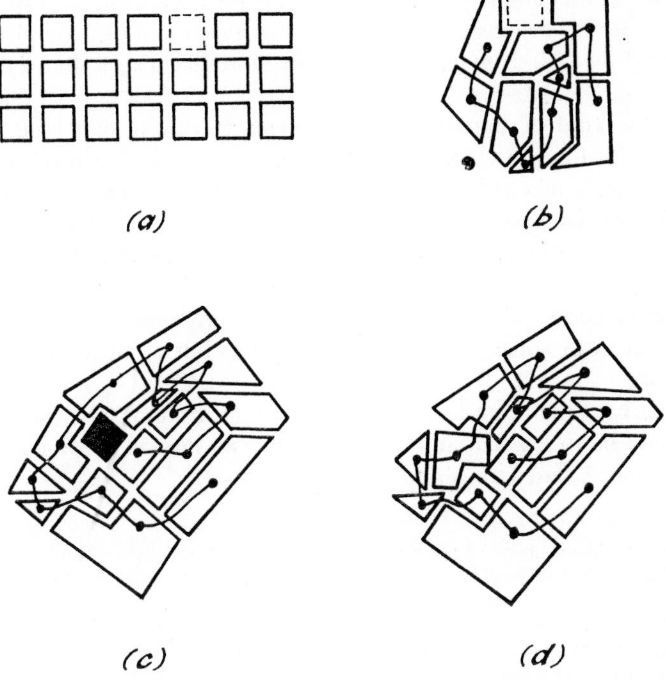

Fig. 14. (*a*) An incomplete crystal tends to select the molecules of which it is composed from its environment. (*b*) Block and string model of a protein indicates an alternative means of selection. (*c*) and (*d*) The specific 'socket' need not be rigid.

protein as an 'incomplete subcrystal' which needs some independent unit (or units) to perfect its subcrystalline arrangement. Like an incomplete crystal, a protein molecule may have a built-in 'hunger' for one or a few particular molecular forms.

We can take the specific socket idea a little further and consider a protein as capable of forming an inverse image of another molecule. By this we mean something more than just a geometrically corresponding slot or hole. There exist in most molecules slight excesses and deficiencies of electrons in different regions (see page 7). A good inverse image of a molecule should have a corresponding distribution of partial charges with their signs reversed (i.e. the image should have partial negative charges where the molecule has partial positive charges and vice versa). There might also be a correspondence between the preferred modes of vibration of the image and the molecule. We need not think of an inverse image, whether a partial one—a slot—or a complete one—a hole—as being necessarily very permanent: it may collapse, more or less, in the absence of the 'completing' unit (Fig. 14*d*). It might sometimes be more like a sock than a shoe, that is.

This kind of thinking about proteins started a very long time ago with Emil Fischer, who suggested, in 1894, the first mechanistic hypothesis of enzyme action—of how enzymes can bring about covalent changes in other molecules. Fischer suggested that a reacting molecule, a 'substrate', fits an enzyme geometrically, as a key fits a lock. This is still the central idea of most modern theories of enzyme action, although we would now be more inclined to say that the fit is in some ways not quite perfect, but strained (Haldane, 1930).

Perhaps the neatest, if still not a quite accurate, way of putting it would be to say that an enzyme contains an inverse image not of a molecule but of a transition state of a reaction.*

---

* An organic chemical reaction is a rearrangement of covalent bonds. A *transition state* represents a critical, difficult, intermediate configuration that the atoms must pass through during this rearrangement. Because a transition state is a strained arrangement, energy, *activation energy*, is required to arrive at it. We say that passing through a transition state involves surmounting an *activation energy barrier*. The situation is similar to shutting a door which has a spring-loaded latch. In changing from 'open' to 'shut' energy is required to overcome the resistance of the spring. Here the transition state corresponds to the point where the spring is most fully compressed: it is a critical state: if you can push the door that far you can shut it.

A detailed X-ray analysis has been made of the enzyme lysozyme (see articles by Phillips, 1966, North, 1966, or Chipman and Sharon, 1969).

Bacterial cell walls are made of networks of molecular threads consisting very largely of modified glucose units joined together. Lysozyme can destroy bacteria by breaking these threads at one particular kind of bond between the glucose units.

The lysozyme molecule has a deep, slightly articulated cleft into which molecules that resemble the bacterial cell-wall 'threads' are found to fit almost, but not quite, exactly. A better fit is achieved if one of the glucose rings is distorted in the kind of way that might be expected for the transition state of a quite plausible bond-breaking reaction. There is a negatively charged protein side group lying close to the carbon atom at one end of the bond which is broken—the transition state expected would have *positive* charge on this atom. Also, there is an acid group on the protein orientated in such a way as to form a hydrogen bond with the oxygen atom at the other end of the interglucose link that is broken by the enzyme—there is thus a ready mechanism for adding a hydrogen ion to this oxygen atom.

ENZYMES AS EXCAVATORS

We can perhaps improve on the idea that enzymes simply 'fit' a transition state by discussing their effects in terms of potential energy surfaces.

A reaction involving a rearrangement of covalent bonds is typically very slow at room temperature. The covalent arrangement of atoms in the glucose molecule, for example, is protected by numerous activation energy barriers corresponding to the numerous covalent reactions of which glucose is in theory capable. These barriers prevent glucose molecules from reacting explosively with oxygen in the air, for example. At high enough temperatures the energy barriers will be overcome, but not usually one at a time: on heating glucose you get a very complicated mixture of products—caramel. In organisms, however, glucose molecules can react rapidly at room temperature, and in different specific ways on different occasions. They may join together in one way to form starch, or in another slightly different way to form cellulose; or, again, the molecules may be progressively broken into pieces, through a quite specific and orderly sequence of processes, to yield energy. Such reactions are directed by enzymes which are able specifically to lower certain activation energy barriers, and so guide the covalent changes along particular routes.

Imagine a clay model of a mountainous landscape including a high level lake. You want to move the water into some particular distant hollow at some lower level. You could do it by violently and randomly skooshing the water about and hoping that some of it would spill into the hollow that you were interested in. But it would be better to excavate the intervening landscape. Enzymes excavate a rather abstract kind of landscape—a multidimensional landscape in which different places correspond to different atomic arrangements, and different altitudes correspond to different potential energies.

Consider the reaction:

$$\text{A—B} \quad \text{C} \rightarrow \text{A} \cdots \text{B} \cdots \text{C} \rightarrow \text{A} \quad \text{B—C}.$$

We make the simplifying assumption that A and C are fixed and that the reaction consists of the movement of B between them. (This assumption is not grossly unrealistic for certain enzyme processes. We make it so as to be able to represent the reaction reasonably completely in just three dimensions.) Now, for A—B to react with C it is necessary not only that the energy of collision should be sufficient to reach a configuration corresponding to the top of the activation energy barrier, but also it may be necessary for the molecules to collide in a more or less specific way: it may be necessary, for example, for C to hit A—B at the 'B' end more or less along the A—B axis. In our simplified set-up it is necessary that B gets a thermal kick not only hard enough, but in the right direction, for B to switch places. So we represent the alternative bonded positions for B by a pair of hollows separated by a saddle-shaped ridge (Fig. 15*a*). There are many different routes between the hollows, but there is only one 'best' route—requiring minimum activation energy—and this passes through the saddle region (B_t). It is the arrangement of atoms corresponding to this place on the landscape that is described as 'the' transition state of the reaction (more strictly it is the transition state of minimum energy). Where reactions involve more variables than in the above simplified example, then multidimensional space is required to describe them; but still there will exist 'saddle regions' corresponding to particular threshold arrangements through which the atoms can most easily pass into alternative configurations. We may suppose that enzymes assist particular reactions by stabilising not necessarily *the* transition state, but *a* transition state in the rearrangement of covalent bonds.

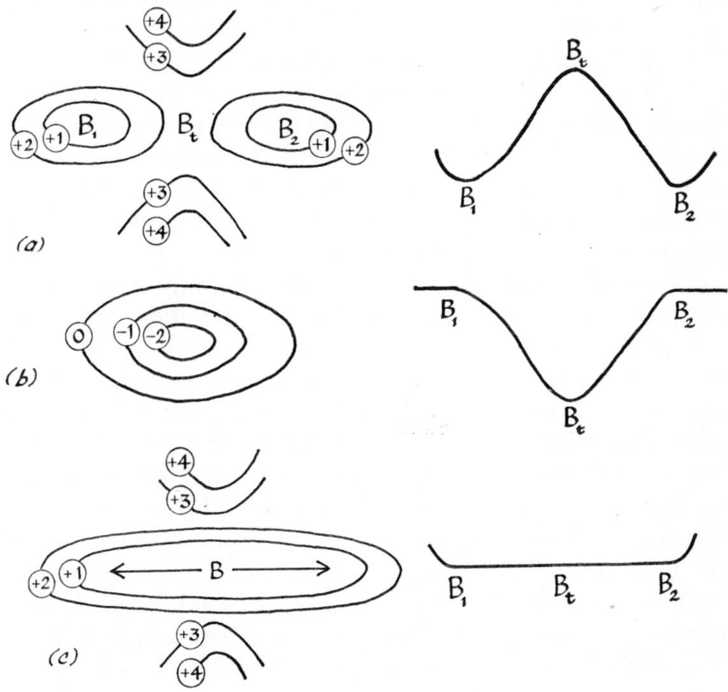

Fig. 15. Contour map (*left*) and cross-section (*right*) of a landscape consisting of (*a*) two hollows separated by a saddle, and (*b*) a single hollow; (*c*) is a superimposition of (*a*) and (*b*), and describes a landscape with level ground between B_1 and B_2.

Now consider another kind of potential energy map: that consisting of a description of the way in which the potential energy of interaction *between* A—B—C and a protein might vary with the arrangement of the A, B and C units. We are now concerned only with external effects of different ways of arranging A, B and C, and not with the internal forces within the ABC system. There need be no relationship between these two maps. For example, it is not necessary that the arrangement most strongly bound to the protein should be the most internally stable one. Indeed, a possible explanation of enzyme action requires a reverse situation, that the binding map has an inverse 'landscaping' to the uncatalysed reaction map. Figure 15*b* represents an 'ideal' binding map for our ABC system. When we now consider the total potential energy for each of the possible arrangements of A, B and C, then we arrive at a map something like Fig. 15*c* in which a route has been neatly excavated between the hollows. (This map is the sum of maps (*a*) and (*b*).)

An engineer might specify the excavation needed to make a road through a ridge by means of a diagram like Fig. 15*b*. It is the map of a hole; it would specify the depth of earth that had to be removed at each place in a landscape such as that described by (*a*) to give a level route such as (*c*). The most economical excavation would be specified by a map with a minimum at the saddle point. Similarly, the most 'economical' enzyme would have a binding map with a minimum at this point corresponding here to the normal transition state. This is what was meant by saying that 'an enzyme binds specifically the transition state of a reaction'; this statement is better than 'an enzyme binds specifically reacting molecules', but it is still only approximate. Indeed, it would be neither necessary nor sufficient, on the general hypothesis which we have been considering, that the binding energy minimum should coincide with the saddle point. The enzyme may choose some route other than the 'easiest' through the ridge separating the hollows. Also, the general shape of the 'hole' defined by the binding map is important. It is no use having a mineshaft drilled at the saddle point. To make a road, your excavation would have to be more extensive than this. So it is with an enzyme. The fitting of the inverse image to a transition state must not be too intolerant: the transition state may be the *best* fit, but the other arrangements along the reaction route should bind also with a carefully graded set of energies—so that the 'earth' is removed in the right places and in the right amounts.

PRECISION ENGINEERING AS THE KEY TO ENZYMIC SUCCESS

It is often remarked that since only a few of the side groups in an enzyme are directly involved in its catalytic action, there seems to be little point in all the other groups. But the idea of a globular protein as being an irregular subcrystal gives point to a large assemblage of constituent units. In the first place, the 'bonds' that maintain much of the subcrystalline organisation are relatively weak, so that many have to co-operate to give a definite coherent structure—just as ordinary crystals that are held together by weak forces only become stable when a fair number of units are assembled. There is a second point which, as we will discuss in chapter 6, may be of great significance for the evolution of proteins. Consider the block and string model in Fig. 14*b*. While it may be true that changes in blocks surrounding the surface pit could be expected to produce the greatest effect on the shape of the pit, changes farther away could also indirectly produce some effect. Such distant changes could act as fine adjustments in the orientation of groups in the crucial region. Now, whatever exact mechanisms are responsible for enzyme action, it seems clear that great precision in the construction of the enzyme is the key, rather than any peculiar kind of force operating between an enzyme and its substrate. Precision of manufacture demands ways of making fine adjustments—and this is perhaps what much of the 'message' in the primary structure of an enzyme is about.

ALLOSTERIC ENZYMES

It is becoming clear, however, that enzymes are often more than highly efficient and specific catalysts. The activity of an enzyme may depend critically on particular molecules present in its environment. For example, a high concentration of the product of a series

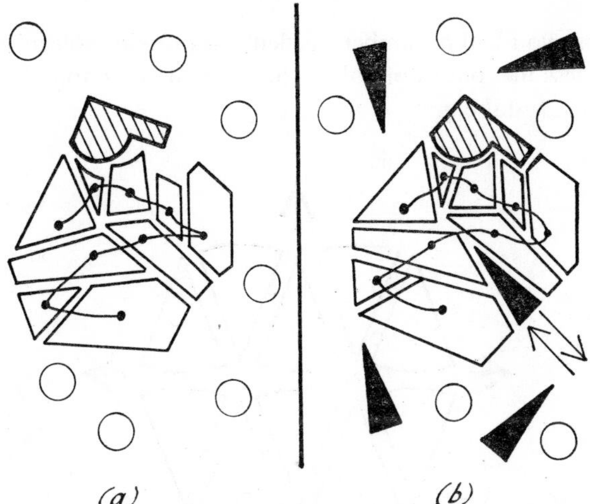

Fig. 16. (*a*) Protein does not fit substrate. (*b*) In the presence of a third molecule the protein may assume an alternative tertiary structure which does fit the substrate.

of enzymic processes can act as a specific inhibitor of the first enzyme in the series. In this way the concentrations of substances within organisms can be controlled both accurately and economically (see Changeaux, 1965). Some enzymes can be even more subtle than this—they may be able to 'decide' whether, or how fast, to catalyse a reaction on the basis of several 'readings' of concentrations of a number of different kinds of molecules in their immediate environment. Monod, Changeaux and Jacob (1963) have suggested that such controlling molecules operate by affecting the tertiary structure of the protein and hence either distort or improve the active catalytic site. They described such enzymes as 'allosteric'. Figure 16 illustrates how units that are not part of a protein primary structure could nevertheless be incorporated in its tertiary structure thus modifying the tertiary structure.

QUATERNARY STRUCTURE OF PROTEINS

Protein molecules may associate into higher order structures by a self-limiting crystallisation process. Haemoglobin, which consists of four myoglobin-like units locked together through crystal forces, is an example of such a *quaternary structure*. There are many other examples among enzymes (see, for example, Cook and Koshland, 1969).

Suppose you had a number of identical protein molecules shaped more or less like pieces of cake. These might come together or give an infinite crystal e.g.

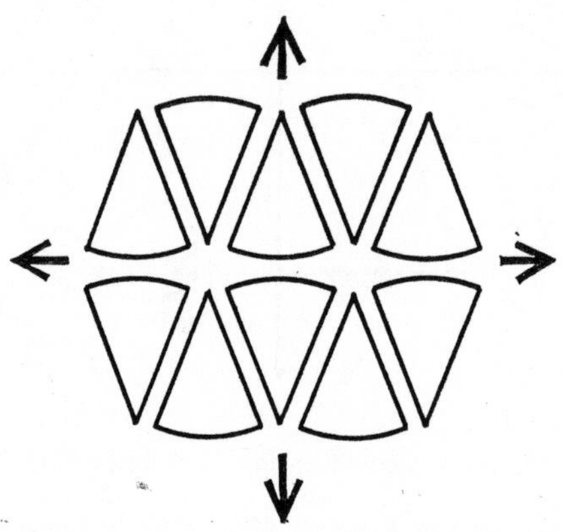

If, however, there exists a complementary fitting between suitable surfaces of the molecule, then some finite object might be preferred, e.g.

Such processes are thought to give rise to the regular protein shells of some simple viruses, like the ones shown in Plate III.

Amino acid sequences can thus specify not only separate globular protein molecules, but also higher order structures by specifying a mode of assembly of individual bricks.*

IS PROTEIN UNIQUE?

If you can talk about parts of an organism as being 'living'—e.g. living tissue, living cell, etc.—then I suppose you may call a protein molecule in an organism a living molecule. You may, if you like, call a gramophone record a piece of musical plastic. But in either case your way of putting it may give the impression that you have missed the point. Any musical properties that a gramophone record may be made to reveal depend on a microscopic pattern which the plastic is able to hold. In a rather similar way the contribution of a protein to a living system depends on its being able to hold quite ordinary atoms in a quite extraordinarily precise and 'deliberate' way. It is not so much the mixed polyamino acid that is living, as the 'music' that is written into it.

Although RNA can have quite an elaborate tertiary structure, the engineering technique embodied in proteins seems to be virtually unique—on the Earth, at the moment. But whether other material

---

* The assembly of more complex structures, for example more complex viruses and some cell components of higher organisms, certainly involves more than a single self-limiting crystallisation (Wood and Edgar, 1967; Laemmli, 1970). But there can be little doubt that the specific fitting together of proteins represents a major factor in the specification of structures at the cellular level (Wolstenholme and O'Conner, 1966).

could equal or improve on the characteristic versatility or functionality of protein, we do not know. We have hardly yet seriously got down to the kind of colloidal engineering that would be required to find out. (See, however, footnote on p. 26.) Protein may well be a makeshift material which was chosen in the first place for opportunist reasons—because amino acids happened to be around during the early stages of evolution (see chapter 8).

Protein is uniquely fascinating to us because it reveals a potentiality of matter which we *might* have otherwise guessed from the study of simple molecules and crystals. But this 'uniqueness' would seem to be both local and temporary.

FURTHER READING

WUNDERLICH, B. 1964. The solid state of polyethylene. *Scientific American*, **211**, 81. REES, D. A. 1967. C.S.P. No. 14 *The shapes of molecules*, Oliver and Boyd, Edinburgh, discuss the conformation of molecules. GREENWOOD, C. T. and MILNE, E. A. 1968. *Natural high polymers* (C.S.P. No. 18). Oliver and Boyd, Edinburgh, discuss structure and function of proteins and nucleic acids. MOSS, D. W. 1968. C.S.P. No. 15 gives an elementary account of enzymes. For an advanced discussion of various proposed general mechanisms of enzyme action see LUMRY, R. 1959. In BOYER, P. D., LARDY, H. and MYRBACK, K. (Eds.), *The enzymes*, 2nd edition Vol. 1. Academic Press, New York, pp. 157–231. WOLSTENHOLME, G. E. W. and O'CONNOR, M. 1966. *Principles of biomolecular organisation*. Churchill, London: This is a collection of papers mainly on the formation of structures within organisms through 'subcrystallisation' processes.

4

Organisms

Whereas most biochemical processes are dominated by proteins, the control of protein synthesis itself is dominated by the nucleic acids. Both kinds of molecule consist of particular linear sequences of units which could in principle be arranged in an enormous number of different ways. These primary structures are thus organised in a way that is analogous to a written message, which may also consist of a linear sequence of a few kinds of different units. We can take this analogy further and regard the primary sequence of a protein as consisting of a message written in a 20-letter alphabet which tells the protein how to 'subcrystallise' and hence specifies its activity.

The base sequences in DNA can similarly be thought of as messages written in a different, four symbol, code. Whereas protein is a specialist in converting a linear sequence of symbols into a functioning three dimensional object, DNA is a specialist in the more formal process of printing. The combination of these two quite different abilities constitutes the basic design feature of all present day terrestrial organisms: replicable DNA messages are translated into functional protein sequences. This may be summarised as a diagram:

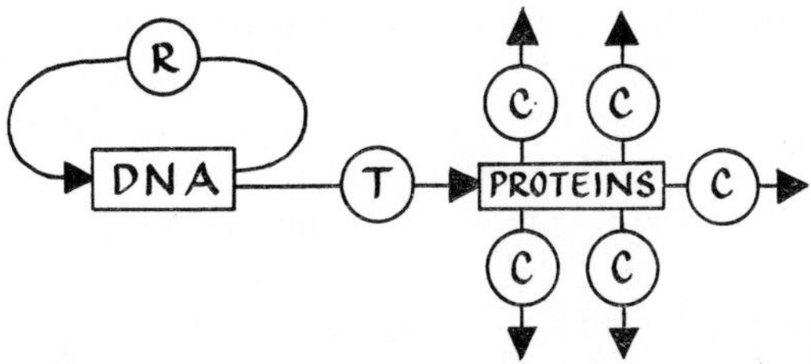

where ®, Ⓣ and © stand for the different kinds of activity: replication, translation, and control. We will return to this particular system later and consider now a more general theory of organisms.

THE GENETIC THEORY OF ORGANISMS

It is only fairly recently that nucleic acid has been identified as the principle, if not the only, replicable structure in modern organisms.

Biologists had long recognised, however, that organisms must have the general *kind* of design summarised above. More than a hundred years ago Mendel explained the inheritance of easily recognisable characteristics between generations of pea plants by supposing that there exist entities, other than the characteristics themselves, which control the formation of the characteristics and which are passed on in the material links between generations—the ovules and pollen. For characteristics to reappear again and again from generation to generation it is necessary that the controlling entities themselves—what were later called *genes*—should be replicable. A total organism is in general regarded as consisting of a set of characteristics, a *phenotype*, arising from a genetic constitution, a *genotype*.

This division allows us to understand in principle not only how organisms can transmit particular traits to their offspring, but how they can have offspring at all. We can think of the genotype as consisting of 'plans' for the construction of the phenotype. Phenotypes may be *reproduced* but only the genotypes are *replicated*. This distinction between reproduction and replication is very important. To be reproduced is simply to be produced again (re-produced). To be replicated is to be reproduced in a particular way; through direct copying processes in which the object itself acts as a template for the production of another similar object. (This may happen in stages, for example through a negative as in photography.) An object derived by replication really is a copy of its 'parent'. This is not necessarily true of reproduction. If offspring resemble their parents, it is not because they are copies of their parents (John's hooked nose was not copied from his father's) but because they arose from similar plans. The characteristics were re-produced: the plans were copied.

Suppose you ask a carpenter to reproduce a piece of furniture. The first thing he must do is to look at the original piece and see what it is made of, find out its dimensions, and so on. He must abstract from the original piece its 'plan'. This he may either keep in his head, or perhaps write down more or less completely on paper. The 'plan' is then duly transferred to new pieces of wood, etc. to complete the job. This kind of non-replicative reproduction requires that the entity being reproduced exists at some point as an 'idea'—something which could in principle be written down on a piece of paper.

Now, how would you go about making a machine which could reproduce *itself*? By analogy with the carpenter, you might start to think that the machine would have to incorporate suitable sensing equipment so that it could look at itself to discover its own plan. But self-reproduction does not require such self-consciousness. The sensing equipment would be unnecessary. You could build in the plans at the start as part of the 'machine'.

The mathematician, von Neumann, demonstrated in the 1940s that a self-reproducing machine was in principle quite possible—and he outlined a general design (Taub, 1963; Moore, 1964). Von Neumann imagined some kind of stockroom containing fairly simple mechanical parts—such as screws, metal plates, wire, and so on. The problem was to invent a machine that could move about such a stockroom selecting the pieces required to make another machine like itself, and then proceed to do so. The centre of von Neumann's design was a set of instructions written, say, on magnetic tape or punched cards, giving an account of how to make the rest of the machine—where to find the parts and how to put them together. The machine would include a manufacturing unit which could follow the instructions and act on them. There is a special point about the instructions themselves, however; they could not be remade by following instructions that were different from themselves. For example, a part of the machine might consist of an instruction card with 86 holes punched in it which caused the manufacturing unit to tighten a screw in the chassis at some point in the manufacture, i.e. the holes produce an effect which in no literal way resembles them. But when it comes to remaking the card itself, if you need another different card to explain to the manufacturing unit where to make the 86 holes, then this different card would also have to be remade involving yet another card and so on. At some point it must be the cards themselves that instruct their own formation, i.e. the cards must be replicated. Only in this way can an infinite regression be avoided. So in addition to a manufacturing unit that can make all kinds of things by *following* instructions, there must be another unit with the more limited task of *copying* them. Figure 17 shows a 'life cycle' for such a hypothetical machine.

Von Neumann's machine solves the problem of 'self-reproduction' in much the same way as it is solved in organisms: by separating the system formally into two parts. One part is completely coded in the form of replicable plans for its construction held in the other part. The machine has a phenotype and a genetic material (which may be cardboard) holding a genotype. If the ability of organisms to reproduce seems to be particularly mysterious, then this arises

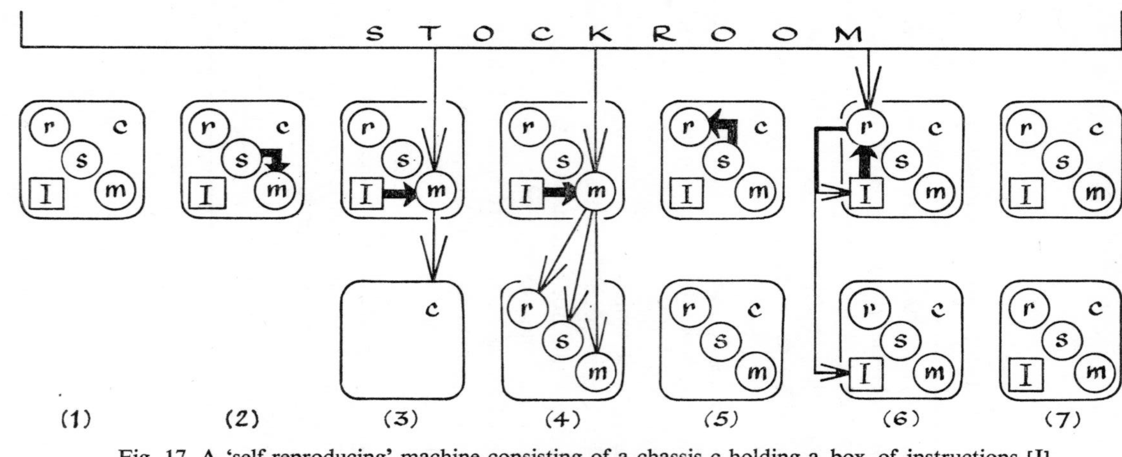

Fig. 17. A 'self-reproducing' machine consisting of a chassis c holding a box of instructions [I] machinery (m) and (r) for acting on and replicating the instructions, and (s) a sequencer or time-switch. (1) Resting phase. (2) Sequencer turns on (m). (3) (m) makes another chassis from materials in the stockroom on instructions drawn from [I]. (4) (m) goes on to make and fix another replicator, sequencer, and another manufacturing unit (m). (This latter is possible because this machinery is being instructed from outside itself.) (5) Sequencer turns off (m) and turns on (r). (6) (r) takes material (e.g. blank cards or magnetic tape) from the stockroom and duplicates [I] adding the final part to a second machine identical to the first one. (7) Resting phase . . .

because in organisms the genes are microscopic: what we *see* when an organism breeds is a succession of phenotypes. Although similar to each other, these have not been copied from each other. The controlling factors are not 'mysterious' because they are extra-material, but because they happen to be invisible to the naked eye. Indeed, the strangeness of living processes generally arises from the minuteness of the machinery on which they depend—and because we continually underrate how elaborate such machinery can be.

WHERE ARE THE GENES?

A higher organism, such as a man or a tree, is not so much like a single von Neumann machine as like a community of such machines working in close collaboration. In present-day organisms it is the cell that is the unit of reproduction.

To a first approximation, a cell consists of a *nucleus* and a *cytoplasm* separated and enclosed by membranes; it reproduces through growth and division:

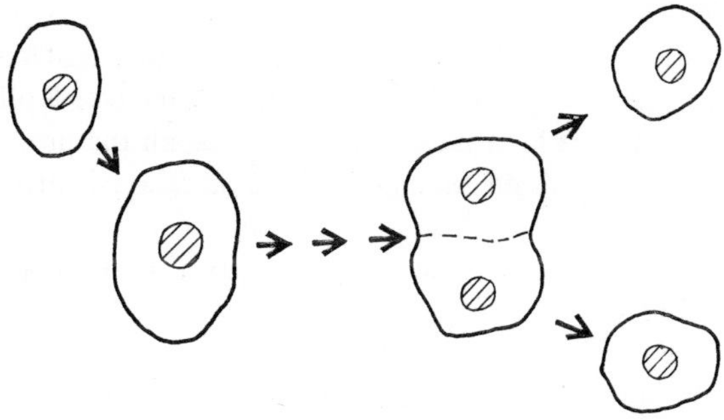

Matter and energy for this process must be obtained from an 'off-balance' environment.

If the nucleus of a cell is suitably examined under the microscope, it may be seen to contain thread-like material (chromatin) that becomes organised into distinct elongated structures (chromosomes) just before the nucleus divides. These become thicker and can be seen as pairs of short cords joined together. The nuclear membrane then dissolves, the cord-pairs separate into individual strands which move to two opposite positions in the cell to form two new nuclei. The cell then divides and the single chromosomes lose their obvious identity, becoming a mass of chromatin. When the chromosomes reappear just before the next cell division, they are again found to be doubled up—and the whole astonishing procedure of *mitosis* can be repeated. The impression is very strong that one is here watching the careful sharing out of newly duplicated plans, one set being given to each of the subsequently forming cells. There is now little doubt that this impression is essentially correct: the genes—or at least most of them—lie on the chromosomes.

It is easy to see why Watson and Crick's suggested secondary structure for DNA, with its implied replication mechanism, created such interest when it was proposed in 1953. Not only does DNA have a likely looking structure for a genetic material, but it occurs in the right place—in the chromosomes almost exclusively. (There was also other independent evidence for DNA as the gene-stuff, but many people still considered that it was the protein in chromosomes that duplicated the genetic messages.)

It has now been well demonstrated that DNA does indeed replicate inside cells (and, with the aid of an enzyme and suitable units, in the test tube) by a mechanism which is formally the one originally proposed.*

Tatum and Beadle (1941, 1945) introduced the idea that a particular gene in an organism specified a particular enzyme (see Koller, 1968, p. 73). The idea was later extended to include all the proteins in an organism. This suggested that DNA sequences are instructions for the synthesis of particular proteins: that the first part of the question, 'How does the genotype make the phenotype?' is 'How does DNA make protein?'

THE GENETIC CODE

)–0*0))0000)0–0**0*)000–)–**)***000*0*****–*–0*–)–0*––*0***000––––000–0***000–)–**)***000–*0**)****–*–0*–)–0*––*0*****– is a coded version of a phrase in the previous paragraph. Since there are more letters in English than symbols in the code, the English letters must be represented by an average of more than one code symbol. One kind of rather simple arrangement—and the above 'message' is like this—is to represent an English letter by a triplet of symbols, e.g. –0*, or)00, and so on. This uniformity, in

* See, for example, Koller, 1968, for an account of these experiments.

having the same number of code symbols per letter, means that there is no need to represent the end of a letter by a special symbol, one can simply start translation by dividing up the code text into triplets from the start of the 'message'. You could, if you like, use quadruplets of symbols to represent letters, but doublets would be inadequate, since there are only 16 ways of arranging pairs of four different kinds of symbols. Triplets give 64 possibilities—quite enough for letters, punctuation marks and 'space'—leaving plenty of triplets over to confuse the enemy by having alternative triplets for the same letter.

Protein is like English, in that it has about 20 different symbols, and DNA is like our four-symbol code. It has turned out that the genetic code is the simple kind of triplet code which we have described. All the triplets are used up by having alternative codings for almost all of the amino acids as well as 'punctuation marks'.

THE TRANSLATING MACHINERY

The idea that DNA sequences are translated into protein sequences would seem to imply that the process, however it is done, should occur in the cell nucleus. This is where most of the DNA in the cell is. In fact, protein synthesis takes place mainly in the cytoplasm in association with small objects, called *ribosomes*, which contain no DNA but contain about equal amounts of protein and a particular kind of RNA known as *ribosomal RNA*. A single ribosome contains two pieces of RNA, one with about 1500 base-units and the other with about 4500. Also in the cytoplasm there are a number of much smaller RNAs each about 77 base units long. These are called *transfer RNAs*. There is yet a third kind of RNA, more variable in length, and altogether more elusive, called *messenger RNA*. Although each of these kinds of RNA is intimately concerned with protein synthesis, their immediate roles in the process are very different: one is a major constituent of an assembly machine, one is an adaptor, and one a message tape.

The transfer RNAs hold the key to the translation process. For each of the amino acids there is at least one particular transfer RNA to which it may become attached covalently. It is thought that each such RNA also contains a particular exposed triplet of unpaired bases. This idea can be represented by a very much simplfied diagram showing one of the transfer RNAs duly 'primed' with its correct amino acid:

For each of the transfer RNAs there is a corresponding specific enzyme which can recognise it and attach the correct amino acid to its tail.

The first stage in the actual translation process, DNA→protein, is hardly more than a transcription of the DNA message onto a differently coloured paper. DNA, with the collaboration of an enzyme, prints off a strand of RNA with a corresponding message, presumably through a mechanism comparable to DNA replication.*

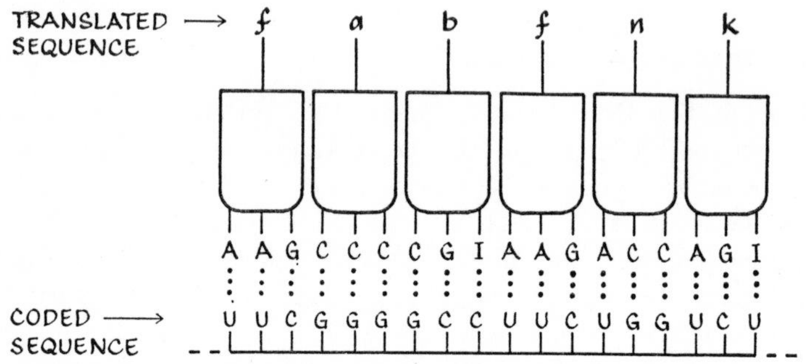

Fig. 18. Highly formalised translation of an RNA 'message' into a protein 'message'. (Transfer RNAs contain a number of modified bases such as inosine (I) which is a modified form of guanine.)

This messenger RNA moves across into the cytoplasm, where its base sequence is translated into a corresponding amino acid sequence through a base-pairing association with primed RNAs. Figure 18 illustrates the idea in a formalised manner. In fact, the process does not occur by such simultaneous lining up of dozens of units but

---

*Two (quite different) mechanisms have been proposed by Florentiev and Ivanov, 1970, and by Riley, 1970.

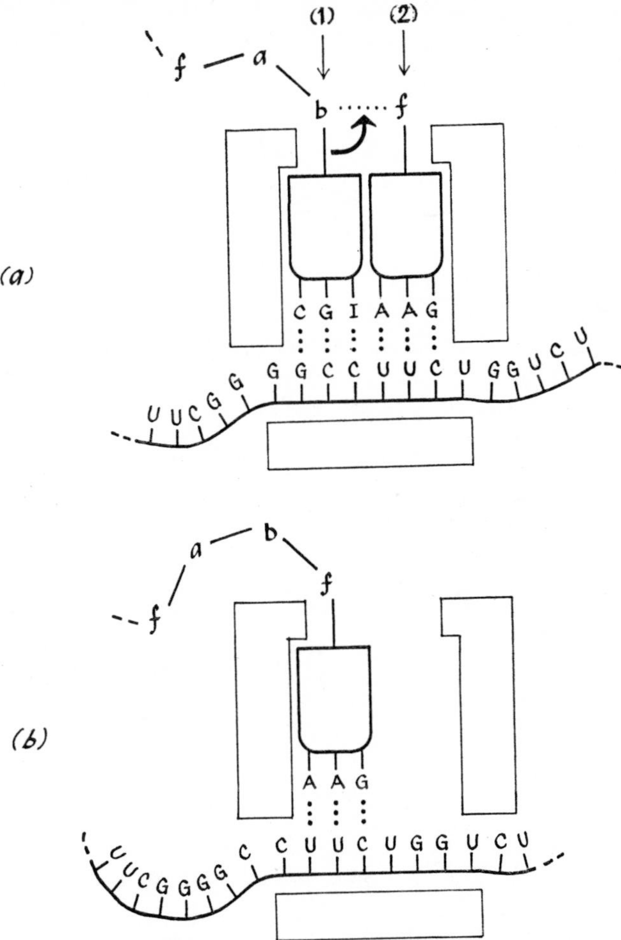

Fig. 19. It is supposed that there are specific slots or grooves in a ribosome into which part of an RNA message tape and two transfer RNAs can fit simultaneously. At stage (*a*) one of the transfer RNAs (position 1) holds the growing protein chain, the other holds a single amino acid that will add next. In this situation the covalent bond holding the growing chain to its transfer RNA switches as shown by the curved arrow. The empty transfer RNA then moves off and the tape advances to bring the transfer RNA which now holds the growing chain into position 1, giving the situation shown in (*b*). Finally, a new primed transfer RNA with an exposed triplet that fits the next triplet of the message 'crystallises' into position 2—and we are back in situation (*a*), except that the protein chain is now one unit longer.

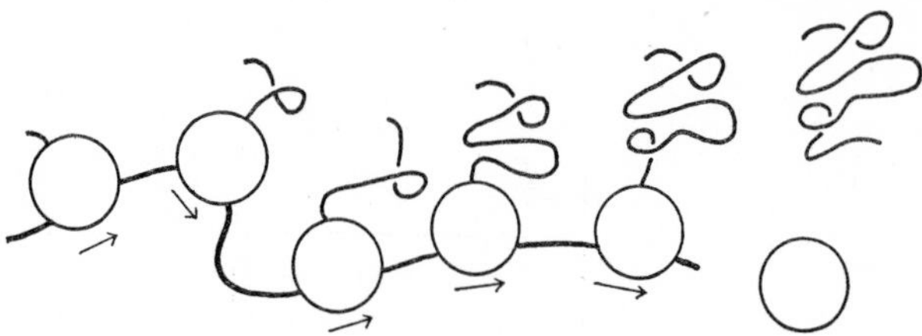

Fig. 20. A cluster of ribosomes 'reading' a piece of messenger RNA.

sequentially with the aid of ribosomes. Figure 19 illustrates, still in a formalised manner, the kind of process that is envisaged.

Clusters of ribosomes (polysomes) have been observed under the electron microscope. These apparently consist of a number of ribosomes simultaneously 'reading' the same message tape (Fig. 20).

For recent reviews on ribosome structure and function see Kurland, 1970; Nomura, 1969; Lipmann, 1969.

'THE MINIMUM PHENOTYPE'

We can improve the diagram given at the beginning of this chapter to include RNA:

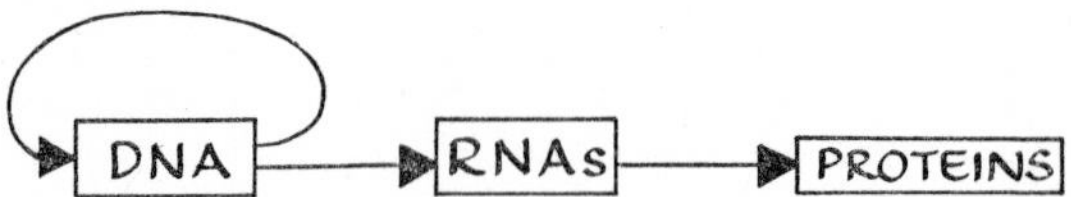

But these processes do not occur *in vacuo*. DNA, RNA, and protein are made out of units which must either be provided by the environment or synthesised by the cell from molecules which are provided. In the latter case whole teams of enzymes may be required. But even if the units are already available in the general environment, the replication of DNA still needs at least one protein, the synthesis of RNA another, and the synthesis of protein needs yet another polymerising enzyme, together with at least a couple of dozen more proteins in the enzymes which prime the transfer RNAs. To make one protein the cell already has to have dozens of proteins. There is nothing immediately illogical in this situation: a factory for making nuts and bolts can be made with the help of nuts and bolts, but it does mean that a cell must inherit more than a book of instructions from its parent. It must inherit also enough pre-formed equipment to read the book. It must inherit a 'minimum phenotype'. A reproducing cell, then, must consist of at least a minimum phenotype together with the instructions required to reproduce it—a minimum genotype. Morowitz (1966) has estimated that for a minimum cell consistent with the current viewpoint of molecular biology one would need at least 45 proteins. Figure 21, then, is a further elaboration of the basic design diagram.

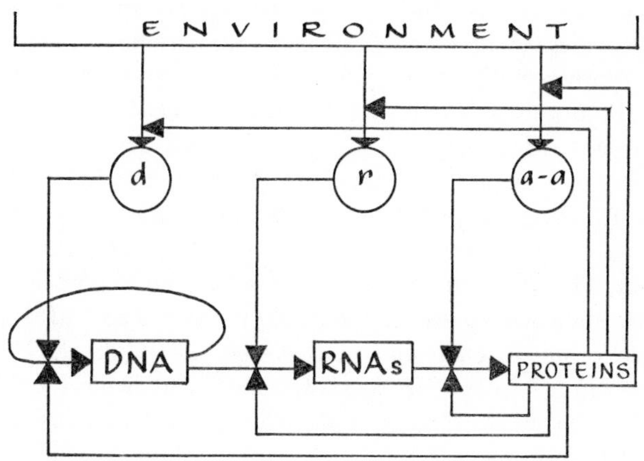

Fig. 21. Present-day terrestrial organisms must have a large number of proteins already made before they can synthesise new proteins.

It is still a gross oversimplification since, among other things, it ignores reactions required to provide energy for the various processes, as well as essential mechanical structures such as cell membranes.

The problem of the origin of life is simply that any conceivable such minimum unit would seem to be necessarily far too complex to have arisen by chance—to have 'nucleated' spontaneously—under any reasonable circumstances. 'Life can only come from life' is no longer a dogma as it was in the immediate post-Pasteur era: but it nevertheless seems that life in fact always does arise this way, and that in nature it must—for any form built on the modern DNA→protein design. Surely there was a radically simpler plan to begin with. What was it like? Von Neumann's 'self-reproducing' machines seem to indicate that quite a complex phenotype, together with a corresponding genotype is essential for *any* reproducing system. Is there a way out? There must be if life really did originate spontaneously as a reasonably probable physico-chemical event during the history of the Earth.

The virus perhaps gives us a clue to the solution of this problem. Some viruses have a phenotype consisting of only one kind of protein which subcrystallises into a box for holding the genetic nucleic acid (Plate III). Yet it can reproduce and evolve. It can do this by living in an environment, the phenotype of another organism, which provides for almost all of its needs; in particular, the complete protein synthesising machinery. All this would be too much to hope for on the primitive Earth, but the example of the virus helps by illustrating that an evolvable organism does not need to have much in the way of a phenotype if its environment is sufficiently obliging. A complex minimum phenotype arises from a kind of mismatching between a genetic material on the one hand and an environment on the other. The origin of life is a problem because DNA seems to be so badly matched to our present environment, and also to any plausible primitive environment that one can think of. The only environment to which DNA seems to be well matched is the inside of a cell. The system works very well (who are we to criticise?) but it does not look like a self-starter. As we will think of it in the final two chapters, the problem of the origin of life is primarily the problem of matching a genetic material—which need not have been anything like DNA—to a plausible primitive environment. We shall try to invent an organism for which the genetic material–environment match is so close that the resulting minimum phenotype is approximately *zero*. We would suggest that the von Neumann machine, like modern organisms, misleads us into thinking that there is a high minimum complexity for *all* organisms by presenting us with a particular genetic material–environment pair (e.g. punched cards–stockroom) which is hopelessly mismatched.

THE GENETIC VIEW OF ORGANISMS

Butler expressed the 'genetic view of organisms' very neatly with the statement, 'The hen is the egg's way of making another egg.' We are accustomed to thinking of an organism as being *really* its phenotype—what we can see, what affects its chances of survival, what immediately impresses us, what makes it, perhaps, beautiful or disgusting or dangerous. The genotype we tend to think of second: it is just the 'plans' for the phenotype—and plans are surely less important than what the plans are about? But it is sometimes as well to consider the 'plans' first: to think of the hen in terms of the egg, to think of the phenotype as a device made by the genotype for its own survival and propagation; to adopt, that is, 'the genetic view'.

A GENE IN A STREAM?

A still more extreme view is to regard the genotype on the one hand, and the total environment on the other, as the only clear components of a living system. The phenotype is then simply a convenient term for describing a rather ill-defined part of the environment where the most intense and persistent genotype–environment interaction is

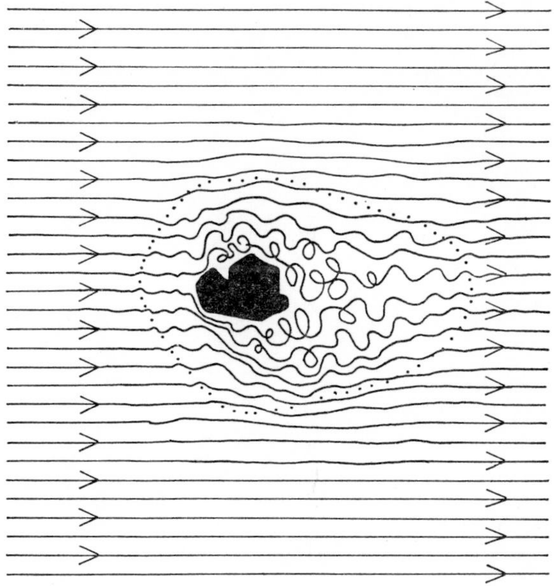

Fig. 22. A stone (compare genetic material) of a given shape (compare genotype) may produce a characteristic dynamic eddy pattern (compare phenotype) when placed in a stream (compare 'off-balance' environment). The eddy pattern is part of the environment—there is no clear boundary between the 'phenotype' and the rest of the environment.

taking place. All the energy and matter in the interaction must be supplied by the environment. The genotype supplies the cunning. It controls the formation of 'eddies' in the environmental stream (extending Sherrington's analogy (p. 10), as a stone of a particular shape might give rise to a particular persistent and dynamic pattern of eddies in a stream in which it was placed (Fig. 22).

The genotype is the finite part of the composite living system. The other part is the rest of the universe—a 'flow environment' (chapter 1). We say that a living system is dynamic and organised: it is dynamic because the *environment* is potentially or actually dynamic; and it is organised because, in the first place, the genotype is organised. This is the essential collaboration.

FURTHER READING

PENROSE, L. S. 1959. *New Biology*, **28,** 92–117, discusses automatic mechanical self-reproduction and gives examples of very ingenious models that can be constructed. See also PENROSE, L. S. 1959. *Scientific American*, **200,** No. 6, 105.

WATSON, J. D. 1965. *The molecular biology of the gene*. Benjamin, New York, combines a genetic and chemical view of the workings of present day terrestrial organisms.

PART TWO — ORGANISATION

5

On Being Organised

It is often supposed that scientific explanations are, or should be, strictly *mechanistic*. A good scientist, so the story goes, should be concerned with relating conditions and events at a particular time to previous conditions and events. There should be no hint of *teleology*, or if there is it must be exposed and apologised for. A teleological explanation is an account in terms of some future state, some 'goal', to which a system is 'striving'. This kind of explanation is regarded as dangerously unscientific; presumably because words like 'goal' and 'striving' contain a strong note of animism. They seem to imply that atoms and molecules are purposeful in the kind of way that we imagine ourselves to be: that they 'see where they are going' and 'decide what to do' and so on.

Certainly the words 'goal' and 'striving' are unfortunate: so are words like 'attraction' and 'repulsion'—yet we manage to use these without implied animism. It is doubtful if even Aristotle, whose science was fundamentally teleological, always thought of processes which he explained teleologically in animistic terms (compare Toulmin and Goodfield, 1965, p. 92). In any case, modern science frequently resorts to a non-animistic teleology—even if she usually keeps quiet about it.* Curiously, in physical science, it is in that most respectable branch—classical thermodynamics—that teleological thinking is most rife.

* Pittendrigh (1958) has introduced the term 'teleonomic' to describe apparently purposeful processes, but for our discussion we do not have to introduce another term provided we recognise that teleology is not necessarily an animistic idea. See Thom, Mayr and Waddington in Waddington (1968) for further discussions of telelogy (teleonomy) in science.

Whether we attempt a mechanistic or teleological explanation may simply be a question of convenience. It may be easier to explain a process in terms of some 'ideal' state to which a system tends rather than in terms of earlier states and events: the 'ideal' state may be much more easily specified.

Suppose that you leave a bucket of water out in the open one night and in the morning you find that it has frozen. You might explain this by appealing to a 'law of nature'—that water freezes when the temperature drops below 0°C—and go on to add that the coldness of the night thus leads to the freezing of the water. This would be a mechanistic explanation, if not a very good one. If you were prepared to consider the phenomenon in greater detail, however, you might well come up with another explanation. You might point out that the characteristic arrangement of water molecules in ice is the best possible compromise between energy and entropy factors: it is the 'ideal' state below 0°C, so, given that the molecules are buzzing about the whole time, trying out various arrangements, they will eventually find this ideal arrangement. Admittedly this explanation does not tell you how long the molecules should take to reach the 'goal': but there is nothing unscientific about it. This teleological explanation is more illuminating than our first mechanistic one.

We may be aware that there exists another kind of mechanistic explanation—a blow-by-blow account of the whole process at a molecular level. But a complete explanation of this kind is only 'better' in theory: it is too detailed to be practical.

Similarly we may think that we could in principle explain the human eye mechanistically; in terms of a long history of the selection of more or less random changes in our distant ancestors. But it still seems worth mentioning that eyes are for seeing with. Indeed, much of the time in biology, as in thermodynamics, it is more practical (and arguably more 'scientific') to take a teleological view.

Biology and thermodynamics are each concerned with organisation, and organisation seems to be a wholely teleological idea. To most people, at least, a purely 'historical' account of how, say, some particular arrangement of atoms arose is not in itself a demonstration that this arrangement is an organised one. Certainly all atomic arrangements have some sort of history, but to be called organised they should also have some sort of function. They should, so to speak, be related to the future as well as to the past. We shall understand an organised system then to be some kind of *functional arrangement of units.**

In practical life there are many kinds of organised systems, and one such system may have different functions from different points of view. A piece of sculpture, for example, is an arrangement of atoms. Its creator may have meant it to be beautiful, or to annoy the critics, or to sell for 2000 guineas. In the event the sculpture may end up by being used as a door-stop or an ashtray. Such considerations stress, not that function is an arbitrary and 'unscientific' concept, but that organisation is not an intrinsic property. Organisation is not so much *in* an arrangement of units as between the arrangement and its surroundings. This relationship may change either through a change in the arrangement or in the surroundings.

Near the surface of the Earth exceedingly numerous arrangements of atoms in time and space are possible. We are strongly inclined to think that some of these, for example the arrangement constituting a cow, are organised in some objective way: that the organisation does not depend on our happening to find a function for it. Have we really any basis for this?

Consider a Martian in his flying saucer in stationary orbit over a golf course. He is investigating the spatial distribution of golf balls on the course. Every 20 minutes he makes a note of the positions of every ball on the course by putting dots on a map. He has no idea of the rules of golf, and has not yet noticed that there is life on Earth, but he would soon appreciate that the golf balls were under the influence of some organising factor(s). He would see that the distribution of dots on his map was very far from random.

Our general basis for suspecting that there are organising factors at work in nature are similar to the Martian's reason for thinking

---

* Whereas in statistical thermodynamics it is important to extend the use of the word 'arrangement' to include the motions of particles (cf. p. 5), in organisms, it looks as if we can, perhaps surprisingly, take a more static view. This is suggested by the phenomenon of cryptobiosis ('suspended animation'). Many quite complex organisms can be 'stopped' by cooling or dehydrating, and then 'restarted' again at some indefinite time in the future by carefully warming up or rehydrating (Hinton and Blum, 1965). The essential organisation must thus be purely spatial and not dynamic. We might guess that this organisation is mainly or entirely in the form of covalently bonded patterns of atoms. (This fits, of course, with the 'gene in a stream' idea (p. 63). Here the essential organisation is the (static) shape of the stone which would re-create at any time the same (dynamic) eddy patterns if put into a suitable stream.)

that there are organising factors at work on golf courses. Of all imaginable arrangements of atoms, some, and some related groups of arrangements, are far more prevalent than others. Some arrangements of atoms are more 'successful' than others in a quite direct and self-demonstrating way—they turn up more often than you would expect on the basis of chance alone.

PREVALENCE AND ORGANISATION

There are just two reasons why a given arrangement of units may be prevalent. It may be prevalent because it forms easily or because it lasts for a long time. Let us say that an arrangement may be more or less *producible* and it may be more or less *persistent*. Figure 23 illustrates these two causes of prevalence.

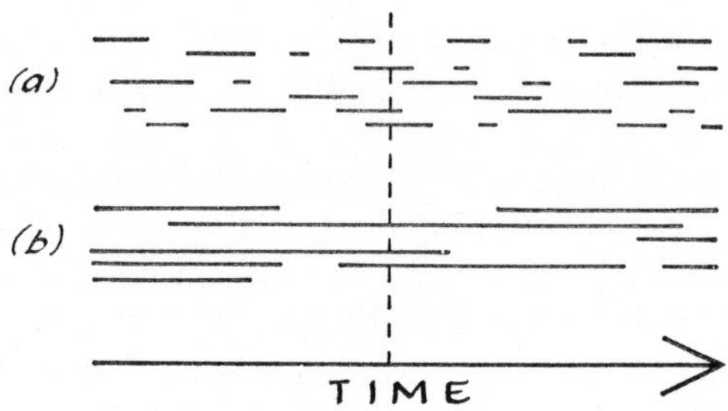

Fig. 23. (*a*) Represents 'life lines' for a highly producible but evanescent structure; at any instant, such as that represented by the vertical line, there are about three of these arrangements in existence. (*b*) Represents life lines for a structure that forms less often but persists longer: its 'prevalence' is also about three.

An arrangement may be highly producible for various reasons. A molecule, for example, may be highly producible partly because it is simple—methane, CH_4, is commoner than any particular isomer of $C_{40}H_{82}$. Or a molecule may be producible because it is easily formed from some other easily produced molecule or molecules; or again because its formation is controlled by another prevalent structure—much as water patterns are produced continuously by stones in a stream. But producibility *per se* is not a *function* of a given particular system. Rather it is an historical fact about that system—

that it *was* easily produced. So although prevalence is self-demonstrating, only one of its causes—persistence—is a self-demonstrating function. The 'point' of naturally organised arrangements of units, then, is continuity in time.

FOUR ASPECTS OF ORGANISATION

We would say that a stable crystal is organised, not because it is very regular, but because the regularity has a function: the crystal is organised because the arrangement of its units tends to lead to its continued existence. If we try to say how *highly* organised a crystal is, however, we are in difficulties. Are we to measure the *efficiency* of the organisation—how stable the arrangement is? Or are we to be impressed mainly by its *complexity*—by how many atoms there are in the repeating unit, perhaps, or by how many words we would need to describe it? Or again, is it the *orderliness*—its precision of arrangement—that we regard as the key feature? Then in addition to being more or less efficient, complex, and orderly, an organisation will be more or less *co-operative*. The critical nucleus in the growth of a crystal from a slightly supercooled liquid, for example, is a strongly co-operative organisation: it will fail totally if cut in half. A big crystal, on the other hand, is not much more organised in this way than the critical nucleus. It can be cut into pieces down to the critical size without producing a complete collapse.

THE INADEQUACY OF NEGENTROPY

Molecular orderliness has the great advantage that it can in principle be measured as 'lack of entropy' ('negentropy'). While it is true that orderliness is necessary for all organised systems, it is not true that very orderly systems are necessarily highly organised in the kind of way that most people would regard as important. It is certainly not true, as is often suggested, that evolution can be regarded as simply a local reversal in the universal drift towards chaos. If this was the most important thing to be said about evolution, then the alleged biblical conversion of Lot's wife into a pillar of salt should be regarded as the most dramatic evolutionary advance of all time. Whatever other difficulties that reaction presents to physical and chemical theory, one thing is fairly clear: the entropy change was negative. The pillar of salt was more orderly than Mrs Lot, but not, we would say, more highly 'evolved'.

Then again, a perfect crystal of a pure substance at absolute zero of temperature normally has an entropy of zero: there is only one micro-state possible. Everything is in its proper place. In a sense its organisation is 'perfect'. But only because the *kind* of organisation is extraordinarily unambitious.

HEREDITARY ORGANISATION

'Survival of the fittest' is a splendid motto for crystal structures. But it does not fit organisms quite so completely. Organisms are not particularly good at surviving: they are not as good as stones, for example (cf. Sherrington, 1940). If organisms are prevalent, it is mainly because they multiply. (A neo-Darwinian motto for them might be something like 'prevalence of the most reproducible'.)

If persistence is the function with respect to which a crystal is organised, what is the function for organisms? Persistence is clearly part of the story: organisms are made to last—if not always, for very long. But if it is their ability to reproduce that is the peculiar cause of their prevalence, it must be in this faculty that the primary biological function lies. Survival (for a time) is simply an inevitable secondary function. And particular techniques of survival, such as the ability to run fast, are subfunctions of the ability to survive. Then again, such an ability will depend on still more remote subfunctions—contraction of muscle fibres, and so on.

Now, here is a problem. We argued that producibility is not a function. Yet organisms seem to be rather good examples of systems which owe their prevalence to being easily formed (think of rabbits). They are highly producible by being reproducible. The way out of this difficulty is to adopt a genetic view: to regard 'the system' as the genotype* rather than the organism and the overall function as the continuity of the genotype through replication. This cause of prevalence is not confined to the genotypes of organisms but to any self-replicating arrangement—e.g. crystal dislocations during crystal growth. It is illustrated in Fig. 24 and is a sort of hybrid between (*a*) and (*b*) of Fig. 23. The individual concrete arrangements do not last very long, but the abstract arrangement itself has an indefinite future as a continuous succession of related types. An abstract

* For sexually reproducing organisms, the gene pool of a population.

arrangement whose prevalence depends on its ability to replicate we will call a *hereditary organisation.* As we shall discuss at some length in the next chapter, an outstanding feature of *some* hereditary organisations is that they may change slowly towards more efficient, more complicated, and more highly unified types—they may *evolve.* This may happen where the pattern which is being replicated creates a specific persistent disturbance in the environment that increases the efficiency of the pattern—i.e. which tends to make the pattern more prevalent through replication. The evolution of a hereditary organisation, that is, may produce a phenotype.

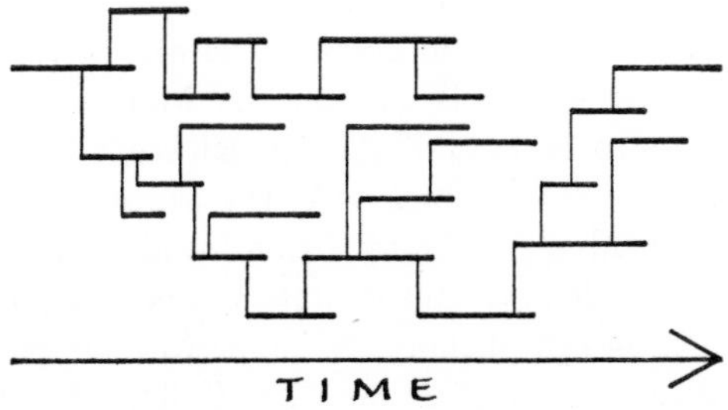

Fig. 24. Replication, indicated by vertical lines, may combine individual evanescence with continuity of the type (cf. Fig. 23).

BIOLOGICAL ORGANISATION

A convenient definition of biological organisation might be formulated from a strictly genetic point of view. One might say that biological organisation is a particular kind of hereditary organisation: one that has acquired somehow the means of developing a phenotype. From this point of view the phenotype is not itself a biologically organised system, it is rather a reflection, or expression of the biological organisation in the genotype. In a similar way one might argue that the undoubted persistence of an eddy pattern in a stream is not due to the physical organisation of the eddy pattern: it is rather a consequence of the physical organisation of the stones, etc. forming the bed of the stream. The persistence of eddy patterns is, so to speak, second-hand. Eddy patterns last because stones last.

Similarly phenotypes persist and reproduce because genotypes persist and replicate.

Now, such a strictly genetic view of biological organisation is useful when discussing the simplest conceivable kinds of organisms—and we will make use of it in the final chapters on the origin of life. But for higher organisms it is rather too simple. In practice biological organisation is a much vaguer idea. We might say that the biological organisation in an organism is that organisation which is common to the genotype and the phenotype of an organism. This is very abstract. What is it that an enormous sequence of adenines, guanines, thymines, and cytosines has in common with the man next door?

You might ask another question. What is it that a succession of black marks on paper has in common with a performance of the play *Hamlet*? The answer is 'an idea'. A Martian might find this idea very difficult to get hold of. After much study he might come to the conclusion that the text of *Hamlet* is a genotype which interacts with its environment in such a way as to bring about its own preferential reprinting. This happens through the development (the theatrical term is 'production') of phenotypes, i.e. performances. The frequent success of performances of *Hamlet* has indirectly determined the long-term success of the text.

We can take this analogy further. A performance of *Hamlet* may be a very elaborate affair, organised in various ways and at different levels. The producer, the stage manager and the actors would each have different ideas as to the *really* important aspect of the organisation of the performance. But the essential organisation constituting *Hamlet* exists completely in the text. In a rather similar way the essential organism—the porcupinity of a porcupine or the oak-treeness of an oak tree—exists completely in the 'text'—the genes. Always the 'text' is much more compact than the 'performance': so if we want to get an idea of how much organisation has been intended by the author or passed on by the ancestors, it is perhaps a good plan to examine the 'text'.

In a man the 'text' consists of the DNA in the nucleus of any of his cells. (This is about 10^{10} base pairs, so you could write the 'text' of a man in a book containing about a million pages. This text would run something like: ACGTTGCAGAGTCGTTGA . . . it would be rather heavy reading.) Somehow this 'text' contains all the information required to make a man.

INFORMATION CAPACITY

The length of a text of any sort gives one a very rough impression of the amount of information that the text *could* be carrying. More precisely, this capacity will depend not only on the total number of symbols in the text, but also in the number of *kinds* of symbols that are being used. DNA, with only four symbols, is rather long-winded. A binary (two-symbol) code would be even more so. By convention the binary representation is taken as a standard in giving a formal measure of information capacity to a symbol sequence. Thus the maximum information capacity, in bits, of a piece of DNA is twice the number of its base pairs, because an equivalent ideal binary representation would be twice as long. (In an equivalent binary representation of a four-symbol code each of the four symbols can be replaced by one of four possible doublets, e.g. A = XX, G = XY, T = YX, C = YY.)

There are a number of ways in which the information carrying capacity of a text will be reduced. Some of these can be seen from an inspection of a text without knowing what it means. For example, if the symbols are used unevenly—as they are in English in which, say, t and e turn up more often than q and j—this must reduce information capacity. One can perhaps see this intuitively by considering an extreme example: if only *one* symbol is used the text could carry no information at all. If you opened a book and found that its text consisted only of a succession of 'a's' you would not bother trying to translate it. But any kind of visible internal regularity has a similar effect. A text that runs ****_*_****_*_****_*_****_*_****_*_****_*_****_*_****_*_**** cannot be saying very much either—and you do not have to be able to read the code to know this. Even the rules of grammar, spelling and syntax greatly reduce the capacity of a language to convey information. These rules could all be inferred by an intelligent Martian with long enough texts of the language at his disposal. The one thing he would not be able to see in the texts *by looking at them in isolation* would be their meaning.

An inspection of a text, then, may tell us something about its organisation: it may define limits of capacity. But even if we knew the complete genotype of a man—if we had the book of a million pages in front of us—we would still find it difficult to assess the actual level of organisation that it represented. We would be in the position of a Martian who could not distinguish *Hamlet* from literary trash—or for that matter from complete nonsense.

We should not, then, expect to be able to judge how 'evolved' an organism is by counting the symbols of its hereditary messages—even although these do contain the essential biological organisation of the organism.

AN IDEALISED MEASURE OF ORGANISATION

In chapter 1 we argued that to hit a golf ball onto a green was to organise the golf ball because we were moving it from a state in which there were a large number of possibilities ('on the golf course') into a state in which there were fewer ('on the green'). We might quantify this. If the area of the golf course is A_c and the area of the green is A_g, then the ratio A_c/A_g gives a measure of how lucky you would be to land on the green by landing a ball at random on the golf course. It is the odds against success by a random approach. It is also a measure of the amount of organising you would have done when you overcome these odds.*

This is a very simple example since everyone agrees that it is a good thing to get a golf ball onto a green: it is not only a special arrangement, but it is a functional arrangement—it makes the golfer happy. (A ball in a bunker conforms to the first but not the second of these requirements.)

Now let us consider another example. Suppose you want to measure the orderliness of organisation of a message. This you cannot do by just looking at the message and counting symbols. Instead you must start by stating explicitly the *function* of the message. You are, let us say, writing a note for the milkman asking him to leave two pints on your doorstep instead of the usual three. As in the 'on the green' and 'off the green' situations, there is a clear-cut criterion of success. If, and only if, when you open the door in the morning there are two pints on the step will your message have performed the required function. Now, there are numerous messages which might be effective. For example, 'TWO PINTS PLEASE' or 'ONE LESS TODAY' or 'KINDLY DEPOSIT TWO STANDARD BOTTLES OF THE PRODUCT OF YOUR COWS'; or even 'DEUX' if your milkman

---

* One might say that the orderliness of the organisation of a ball on a green R_0 is given by: $R_0 = \log A_c/A_g$ (cf. p. 9).

is quick on the uptake, speaks French, and is not easily offended. In principle you could perform an experiment to find out how much organisation there is in a particular successful message in the particular circumstances. You would leave out a different note each night until you had tried out every possible sequence of letters and spaces of the length of your message—this would include also all possible shorter sequences. You would then count how many of these produced the required result. In 'TWO PINTS PLEASE' there are 27^{16} possible sequences (i.e. 10^{23}) so the experiment for sequences of this length would take rather a long time. But given the patience and longevity, you might find that, say, 83 sequences out of these did the trick. Then the ratio $27^{16}/83$ would be a measure of the orderliness of organisation of the message.

This may sound like a distinctly unrealistic experiment, but, as we shall see in the next chapter, it is not so unlike the way in which nature tests a protein sequence to decide whether it is suitably organised or not. But nature never has to measure the orderliness of her products. If a protein sequence works well, or if a genotype creates an effective phenotype, that is good enough. We should, I think, take a similar attitude. To measure the orderliness of the biological organisation in, say, a man we would have to embark on a milkman-type experiment. First we would have to define 'man': then we would have to try out DNA sequences to find what proportion of all possible sequences would indeed produce a man by directing a suitable development process. You would have to take a representative sample of the $4^{10^{10}}$, i.e. $10^{6000\,000\,000}$, possible million page sequences. It is difficult to say how big this sample would have to be, but I would be surprised if $10^{100\,000}$ trials produced a single success. The experiment hardly seems worth the trouble.

COMPLEXITY OF GENOTYPE AND PHENOTYPE

Instead of trying to measure amount of biological organisation in the sense of its orderliness, one may try instead to measure complexity. As a first attempt to measure the complexity of any object one can associate an 'information content' with it. This is the number of bits of information that would be required to specify the object with an agreed precision using agreed 'bricks', e.g. atoms. This measure of the complexity of an object, then, is the minimum length of a binary message that would specify the object. There are, however, two ways in which you can specify an object and these can lead to quite different estimates of 'information content'.

You might simply describe the object, brick-by-brick, giving a 'blueprint' in which every detail of the object is symbolically represented. The most compact blueprint of this kind would provide a measure of the 'amount of effective structural detail' in the object. Thus by a 'highly detailed' object we would mean that a message describing the object, brick-by-brick, would have to be very long: either because the units are very numerous; or because they can be of a large number of different kinds; or because they can be in a large number of significantly different positions—or perhaps for all of these reasons.

It follows from the above that a brick-by-brick blueprint of an object must itself be at least as detailed as the object. If the object is being specified to atomic dimensions, and if its blueprint is made from a similarly sized 'alphabet' of atoms arranged with similar precision, then the blueprint must also be at least as *big* as the object. The genotype of an organism, then, cannot contain such a brick-by-brick blueprint of the phenotype.

Another way of specifying an object is to specify a process or procedure that will give rise to it. Such a specification is an *algorithm*. For an object consisting of an arbitrary arrangement of units, the specification of a process that will give rise to it will *usually* be more detailed than a simple brick-by-brick specification. But the units constituting most objects in the world are so arranged that the specification does not have to be so long-winded as for an arbitrary arrangement. In particular this is true of crystals and of phenotypes. Let us consider crystals first.

Suppose that you want to specify a particular crystal. The atom-by-atom approach would produce a 'message' that was very lengthy—and indeed very boring because it would be highly repetitive. But there is a much easier and more sensible way of specifying a crystal—particularly an ideal one. You could specify how to put together the atoms constituting the repeating pattern and then add something to the effect: 'Continue like this until the dimensions of the crystal are such and such.' In this way there can be far less detail in the specification than in the object being specified.

Similarly, the 'object', 333333333333333333333333333333333333
33333333333333333333333333333333333333333333333333333333333
33333333333333333333333333333333333333333333333333333333333

3333333333333333333333333333333333333333, can be specified concisely by specifying a way of producing it: 'Write "3" 198 times.' We will use the word 'development' to signify the working out of an algorithm. Quite a simple kind of machine could develop the above algorithm, i.e. obey the instruction, 'Write "3" 198 times.'

We can regard the growth of a crystal as a development. Here the algorithm is the seed crystal containing a suitable dislocation. No special machinery is required for the instructions to be carried out, only an environment that is suitably off-balance, i.e. supersaturated.

Highly regular objects, then, can be specified by 'compact' algorithms. But what about this sequence: 17108972589718074143952830793629902608076554555453245818343255130543516432376912466379191111965786082205036734049564234861371774961138104459104482535212494659899522507940259887336645113104023424013049368985267957359091851929066647636392705738600295487428650940053510353852459639474359553172800164308378394874578195621283691115658708500040781396853030778257813849856692950471963508932801857372575553419411939681323335748770973750927141300732417102035051697754984343561187933295519151457453789138048055187827977590775000785579513981749607827046276161312517742057997170554688538703689036095806399241086011592997020790226888203087101533653915806041722653430037764243465142432560124591703100088643979486942002854170097571338930915447098888372333024657251637441276280296188483408232272319501403895185152063432262261261612431271509190879459978732133255390601413833379281814639023615443036233836886179885626005055280120422598617062894203619648023806809643832836096? You might think that there is no compact way of specifying such an irregular 'object'. But you would be wrong. The above sequence can be specified like this: $477! \div 10^{117}$. This is an instruction to carry out the following operation: 'Multiply together all the integers from 1 to 477 and divide by 10^{117}.' Again a machine could carry out this operation, i.e. develop the algorithm.

It is clear, then, that it is not *only* such 'objects' that have some obvious regularity about them that can be compactly specified. A development process can give rise to a structure which is not only bigger, but more complicated than its specification. (An extreme example would be a 'black box' programmed to evaluate $\sqrt{2}$. Such a 'black box' would be a finite entity of finite complexity; yet it could generate a unique and infinitely complicated result.) If this seems paradoxical, one must remember that the number of possible complicated objects which can be compactly specified is limited by the number of possible specifications, and this is limited by the amount of detail in these specifications. Only *some* complicated objects have a compact specification.*

Now can we think of a physical self-operating algorithm like the crystal seed, but which could give rise to a structure more complicated than itself? We want the physical equivalent of $477! \div 10^{117}$. What about a stone? It will create an eddy pattern if put in a stream. The simplest way of specifying the eddy pattern is not by describing in detail the way the water is moving but by specifying the shape of the stone and the general characteristics of the stream into which it should be put.

Dancoff and Quastler (1953) tried to estimate very roughly the 'information' in an adult man by estimating the amount of information that would be required to specify the molecule-by-molecule construction of an exact copy. They supposed that there are roughly 500 different kinds of molecular units out of which a man is, in effect, made and that one can think of each as having twelve possible orientations. This leads to about 7 bits for specifying the orientation and type of one molecule. Given that there are about 10^{25} molecules in all, they arrived at the figure of 5×10^{25} bits in a man after allowing for featureless aspects such as water. Using atoms rather than molecules as bricks, they arrived at the figure of 2×10^{28} bits. A similar discussion of the zygote from which a man develops (i.e. the cell formed by the fusion of sperm and egg-cell) leads to an estimate of about 10^{11} bits. As Apter and Wolpert (1965) point out, however, this brick-by-brick mode of specification is quite un-

---

* The word 'complicated' is difficult to define precisely. We did not regard the row of 3's as a complicated structure because a simple way of describing it is so obvious. But between such a simple example and other very difficult ones like the development of $477! \div 10^{117}$ there are examples which might be obvious to some people and not to others. In a general way one might say that an object is complicated if it is difficult to see an easy way of describing it. Clearly this is very subjective, but perhaps we could agree that repetitiveness is easy to see. In that case we could say that the complexity of an object is the minimum number of bits of information that would be required to specify it by a brick-by-brick approach allowing short cuts through the use of algorithms to specify repetitions. (Compare 'detailedness' which allows no short cuts: hence a big crystal is more detailed but not more complicated than a small one.)

realistic, and tends to lead one to the conclusion that because the adult is more complicated than the zygote then some supernatural agency must be at work during development to supply the extra complication. It is more likely that phenotypes belong to that class of relatively complicated objects that can be relatively easily specified.

There are interesting consequences of the idea that genotypes are algorithms like $477! \div 10^{117}$, or that they are like a stone which produces a particular eddy pattern. You would not, for example, expect to find a correlation between the complexity of genotype and phenotype. Even if you could take account of repetitions, randomness and other factors reducing the information capacity of the genotype; then you would still not expect to be able to correlate this *essential information* with the complexity of the phenotype. Suppose that a simple genotype happens to produce rather an elaborate phenotype. If this phenotype happened to be 'just the thing' for some particular biological niche, then it would be successful—and good luck to it. But on the whole this would be unlikely. The chances are that complicated phenotypes that just happen to be very easy to make will not be particularly efficient. Now, natural selection tends to increase the efficiency of the phenotype. In the above case this might well be achieved by simplifying it. Suppose, however, that there are no easy manufacturing routes to simpler and more efficient forms. More complicated messages might then be evolved. Using the numerical analogy, the 'phenotype' 1710897258971807414395283079362990260807655455545324581834325513054351643237691246637919111196578608220503673404956423486137177496113810445910448253521249465989952250794025988733664511310402343401304936898526795735909185192906664763639270573860029548742865094005351035385245963947435955317280016430837839487457819562128369111565870850004078139685303077825781384985669295047196350893280185737257555341941193968132333574877097375092714130073241710203505169775498434356118793329551915145745378913804805518782797759077500078557951398174960782704627616131251774205799717055468853870368903609580639924108601159299702079022688820308710153365391580604172265343003776424346514243256012459170310008864397948694200285417009757133893091544709888837233302465725163744127628029618848340823227231950140389518515206343226226126161243127150919087945997873213325539060141383337928181463902361544303623383688617988562600505528012042259861706289420361964802380680964383283609 6 can be specified by the simple 'genotype' $477! \div 10^{117}$. But let us suppose that this 'phenotype' is too fancy: something plainer would be more efficient. Say the same sequence but with the last 20 digits all 1's. This simpler 'phenotype' could be achieved through the evolution of a more complex 'genotype', viz; $(477! \div 10^{117}) - 12695698532721724985$.

RECAPITULATION

We have considered two general propositions about organisation. In the first chapter we considered the idea as it is understood in statistical thermodynamics. A system is organised if it is in a macro-state for which relatively few micro-states are possible. We can generalise this and say that an organised arrangement is one that belongs to a sub-set of possibilities. Thus your golf ball is organised when it it on the green, and even more when it is in the hole because, among other things, 'on the green' is a sub-set of 'on the golf course' and 'in the hole' is an even smaller sub-set.

In this chapter we argued that another proposition is in reasonable accord with the general use of the word 'organisation'. This second proposition is that organisation is the functional arrangement of units. This specified broadly the kind of sub-set to which organised systems belong. Combining the two propositions, then, we have that *an organised arrangement belongs to a functional sub-set of possible arrangements of units.* (This means, for example, that you were only really organising that golf ball if you *meant* it to land on the green: if it fulfilled a purpose.)

The next stage was to identify functions that are independent of an external judgement—that are 'self-demonstrating' or 'natural'. At root there is only one—continuity in time. A very simple example is a stable molecule. But the organisation of a system is the arrangement as such: it need not necessarily be associated with a particular collection of atoms. An organisation may be able to persist through replication: it may, so to speak, be transferrable to other collections of atoms. It is this form of continuity that characterises hereditary organisation and which is the basis of life.

We have noted also that a given primary organisation may sometimes give rise to derived effects which can be dynamic, complex, and yet persistent. A stone may, in a suitable flow environment, give rise to eddy patterns. In the first place these patterns are simply *consequences* of the primary organisation. But there is a further possible twist. The derived effects may contribute to the primary function. One might conceive of eddy patterns that in some way stabilise the stone that produces them. In such a case the derived effects would become part of the organisation. It is doubtful whether eddy patterns ever do much good for stones, but phenotypes are certainly useful to genotypes: they contribute to the hereditary function. It is characteristic of biological organisation that it thus exists in two forms within a genotype-phenotype system.

As we shall discuss in the next chapter, such a genotype-phenotype system within a variable environment tends to change in a very curious way: the originally single function of continuity through replication remains, but slowly as generation succeeds generation the means to this end may become increasingly subdivided with the overall function becoming a compound of diverse subfunctions that are critically interdependent—this is 'life'.

6

On Becoming Organised

A consortium of brewers might arrange the construction of the 18th hole of a golf course in such a way that even the most incompetent golfer could not fail to reach the hole in one shot. One design would be a large conical crater with the hole at the base and the tee somewhere on the inner wall at least 400 yards from the rim. With this arrangement no one could ever drive out of the crater, and provided the grass was kept well cut and the sides of the crater were steep enough, success would be guaranteed—even for a random golfer. The brewers, you might say, would have prearranged the outcome. Or you might say that they had set up a 'self-organising' situation. In a long history of such a golf course you could be fairly sure that no two golfers would make exactly the same drive off the 18th tee in terms of initial speed and direction. The number of initial possibilities would be very large. But the number of final possibilities for such a drive would be small—the ball would end up in the hole. Given that 'in the hole' is a possibility which fulfils some sort of function, then the process could be described as 'self-organising'. Without bringing in the question of function, we could simply describe such a process as a *convergent* process, i.e. one which proceeds from a larger set of possibilities into a smaller set. In general, the behaviour of a ball placed anywhere in a uniform surface depression is a model of a simple convergent process. (Contrast the behaviour of a ball placed on a mound, or the difficulty of playing onto a green at the top of a mound.)

Teleological discussions are particularly apt in considering simple convergent processes, since the goal is distinct and easily defined, and the exact initial conditions and routes taken are not very important. If teleological discussions are frequently apt for biological processes, then this is perhaps because these are often convergent—at least in part: they can be modelled—in part—by the behaviour of a ball in a basin with a single lowest point. If teleological discussions are nevertheless inadequate for a complete account of biological processes then this is perhaps because the proper model surfaces are not as simple as this: they are complicated 'landscapes' containing not only hollows but rises, ridges, valleys, saddles, and so on.

Waddington (1957) has discussed biological development—the progressive changes in an individual from conception to the grave—in terms of an 'epigenetic landscape'. Think of a gently sloping plane with branching grooves in it (see Fig. 25). Placed in a suitable

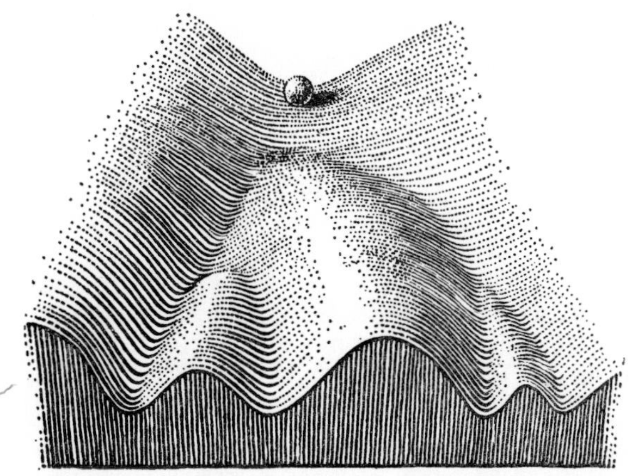

Fig. 25. Part of an epigenetic landscape. The path followed by a ball as it rolls down towards the spectator corresponds to the developmental history of a particular part of the egg. (Reproduced from Waddington, *The strategy of the genes*, 1957, by permission of George Allen and Unwin Ltd.)

'catchment area' near the top, a ball would roll down and arrive at one of a few more or less distinct possible places. This corresponds to the rather uniform looking initial cell of, say, a man producing a variety of distinctly different results (e.g. producing muscle cells, nerve cells, kidney cells, and so on but not intermediate types). Development is a 'self-organising' process, like play on the 'brewers' 18th', only here the landscape is contrived by the genes (see Fig. 26). The genes thus greatly prejudice, even if they do not completely determine, the *course* of development. (Contrast the golf analogy where only the final outcome was 'built in'.) The genes create what

Waddington describes as *chreods*, i.e. stabilised preferred pathways of change, rather than a distinct single goal or telos.*

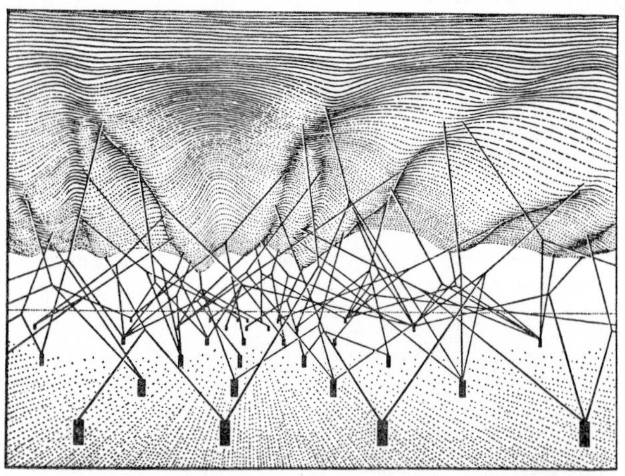

Fig. 26. The complex system of interactions underlying the epigenetic landscape. The pegs in the ground represent genes; the strings leading from them are the chemical tendencies which the genes produce. The modelling of the epigenetic landscape, which slopes down from above one's head towards the distance, is controlled by the pull of these numerous guy-ropes which are ultimately anchored to the genes. (Reproduced from Waddington, *The strategy of the genes*, 1957, by permission of George Allen and Unwin Ltd.)

'CREATIVE' PROCESSES

By a creative process we mean among other things a process that is in some way desirable or interesting, but in which the desirable or interesting elements have not been strictly predetermined. Improvisation is creative. A performance of a set score is less so. Playing a

* Like the landscape which we used to discuss enzyme action (Fig. 15) the epigenetic landscape is really a multidimensional diagram: the course of development of part of an organism is stabilised not only with respect to a single variable, like length, but with respect to numerous variables corresponding to different conceivable chemical constitutions, details of structural forms and so on. A groove can only represent the stabilisation of one variable; a chreod is a multidimensional 'groove' stabilised with respect to different variables in different ways.

gramophone record is not at all creative. Biological development is like a performance of a set score. The composer does not specify completely an exact performance, but he hopes greatly to limit possibilities. In generating a performance one feels that it was the composer that did most of the organising. He had a wider field of possibilities to explore.

In this chapter we will be considering mainly evolution. This seems to be the outstanding example of a 'natural' creative process. We can see it as a writing of scores—as an elaboration of DNA sequences. Yet there is still room to argue about how creative a process evolution is. The composer may have more freedom than the performer: evolution may be less predictable than development, but in none of these cases is the freedom quite untrammelled. Waddington's chroeds are the trammels of development which specify a set of possible processes. In discussing evolution we seek, among other things, to get some idea of the inbuilt restraining factors. We will come to discuss these later in this chapter in terms of a 'landscape'. This is a much wider and wilder country than the epigenetic landscape. It contains no made-up paths, but its contours limit possible or likely routes. If we say that evolution is both a natural and a strongly creative process, then we mean that chance plays a crucial part in choosing the routes. It is still possible that there is no such thing as chance, that every event has a complete causal history, that the present was always completely written in the past: that, for example, the primitive fireball, which is supposed by some to have been the initial state of the universe, was a kind of cosmic zygote: that there was no originality after the formation of that zygote: that everything from then on was an inevitable working out. But in practice, at least, there *seem* to be chance events that often seem to exert a crucial control. In particular, evolution *seems* to be an adventure, i.e. it has no prearranged outcome.

In trying to get into perspective the role of chance in evolution, we will start with a now classic example of the incompetence of 'pure' chance as a creator.

HOW TO GET A MONKEY TO WRITE *War and Peace*

Eddington remarked that an army of monkeys strumming on typewriters *might* write all the books in the British Museum—although he did not recommend the experiment. But we must admit that given long enough the monkeys would write any text you care to specify. You would need someone to recognise when they had done their work (books, like plays, are only 'self-selecting' in a suitable intellectual environment) and just how long the monkeys would be expected to take would depend on exactly how the selection was done.

Consider a monkey hitting one key per second at random on a 30-key typewriter. How long would it take to hit on the sentence, 'WELL, PRINCE, SO GENOA AND LUCCA ARE NOW JUST FAMILY ESTATES OF THE BUONAPARTES'?

On average the monkey will produce any given symbol twice a minute, so if you selected each symbol individually as it appeared in the correct order it would take about 40 minutes for the job to be complete. Suppose, however, that you regard a sequence as being selectable only when it constitutes a complete word (or space or punctuation mark). The monkey must chance on 'WELL' before it can proceed, and then on 'comma,' and then on 'PRINCE' and so on. About how long would it take to finish the sentence with this trial and selection procedure?

Consider the first word 'WELL'. The monkey would hit on a 'W' about once every 30 seconds. Most of these 'successes' however would not lead anywhere since usually the 'W' would not be followed by the necessary 'E'. Such a double piece of luck could only be expected on about 1 in 30 of the occasions that 'W' had turned up. On average, then, you would have to wait 30 times as long for the 'doublet' as for the single letter, i.e. 30×30 seconds. But still only 1 in 30 of these promising doublets would in fact happen to be followed by the correct next letter 'L', and similarly only one in 30 of the suitable triplets would in fact 'mature' into the complete word. So you would expect to have to wait for $30 \times 30 \times 30 \times 30$ seconds, i.e. about 10 days using such 'word-by-word selection' rather than about $30+30+30+30$ seconds $=$ 2 minutes, which you would expect for 'letter by letter selection'. Table 1 gives the average times of waiting for up to 12-letter words using these alternative procedures. As can be seen from the table, the sentence, 'WELL, PRINCE, SO GENOA AND LUCCA ARE NOW JUST FAMILY ESTATES OF THE BUONAPARTES' would take about 600 000 000 years—almost all of this time due to the last word with its 11 letters. Up to this word the expected waiting time would have been about 750 years, and most of *this* time would probably have been spent in waiting (about 700 years) for the

7 letter 'estates' to turn up. So the moral seems to be this: if you want to build up an organised system by random trials followed by selection, use many small stages rather than a few large ones: adopt a letter-by-letter approach rather than a word-by-word approach.

TABLE 1. *The times that you would expect to wait for a monkey to type out a word of up to* 12 *letters long using two different procedures* (*see text*).

Number of letters in word	Expected waiting time							
	Letter-by-letter selection		Word-by-word selection					Millions of years
	Sec.	Min.	Sec.	Min.	Hours	Days	Years	
1	30	½	30	½				
2	2×30	1	30^2	15				
3	3×30	1½	30^3		7½			
4	4×30	2	30^4			9½		
5	5×30	2½	30^5			280		
6	6×30	3	30^6				23	
7	7×30	3½	30^7				700	
8	8×30	4	30^8				21 000	
9	9×30	4½	30^9					2/3
10	10×30	5	30^{10}					19
11	11×30	5½	30^{11}					560
12	12×30	6	30^{12}					17 000

And do not even contemplate a sentence-by-sentence approach. Yet, and here is the apparent paradox, in organisms it seems to be the sentences, paragraphs, and even books that are meaningful. An enzyme is more or less a whole protein molecule. At a higher level it is the combination of countless individually meaningless molecular units that make up an eye. Even an eye is no use, though, without a brain; a brain without a circulation and so on. It is quite characteristic of biological organisation that it is highly co-operative, and it is just such 'all at onceness' that chance seems to be so incompetent at producing. How do we escape from this difficulty?

Consider again the growth of a crystal from a supercooled liquid. The first stage, the appearance of a suitable nucleus, is notoriously unpunctual. And we know why: because this is a 'word by word' selection. A fair number of units have to come together in the right kind of way against the odds. From then on, however, selection is 'letter-by-letter', growth proceeds by the selective accretion of new molecules more or less one at a time. It is as if the monkey had to get the whole of the first word before it could begin to write *War and Peace* on a letter-by-letter basis. (It would take, then, about 10 days for the first word, about 35 minutes for the first sentence, and about another 3 years for the whole book.) The creation of an individual organism is similarly a double process. Here there are two key questions. How did the genotype arise? How did the phenotype develop?

The main *difference* between the formation of a simple crystal and a present-day organism is in the degree of organisation of the initial 'seed' and in the way in which that organisation arose—we will return to this. The main *similarity* would seem to be in the nature of the ultimate molecular processes through which the initial 'seed' develops. Here, in each case, it looks as if 'letter-by-letter' trial and selection processes are of pre-eminent importance. On the growing crystal, individual molecules are continually 'trying out' different positions and those which are most stable are most likely to be retained. In the developing organism there are similar trial and selection processes. Consider just one stage during the synthesis of an enzyme molecule. We might ask, 'How does a correct primed transfer RNA know to come to the ribosome so that it will key with the next three bases in the messenger RNA?' (See Fig. 19). Of course the answer is that it does not know. All the possible 'letters' (primed RNAs) are around and in haphazard motion: all the possibilities can thus be tried out. Only the letter that fits is accepted. What appears to be a smooth goal-directed activity on the large scale turns out to be a jumpy trial and selection procedure when seen at a microscopic level. This, like other biomolecular mechanisms, works because the number of possibilities on trial at each stage is not too great—it is a 'letter-by-letter' process. The folding of a protein is another example (p. 39). The genes hold the overall plan. They put it into effect by making use of the tendency for matter to organise itself, through 'letter-by-letter' trial and selection at the molecular level.

It is one thing to become organised by following an overall plan: it is quite another to form the plan in the first place. Individual molecules in an organism, or a growing crystal, if not quite

(a)

(b)

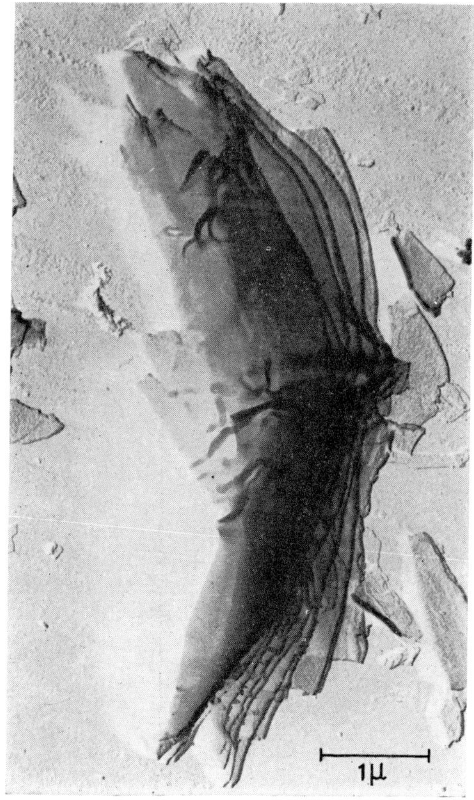

Plate I. Electron micrographs of polythene crystals showing layered structure. 1μ = a thousandth of a millimetre. (From Keller, 1964.)

(a)

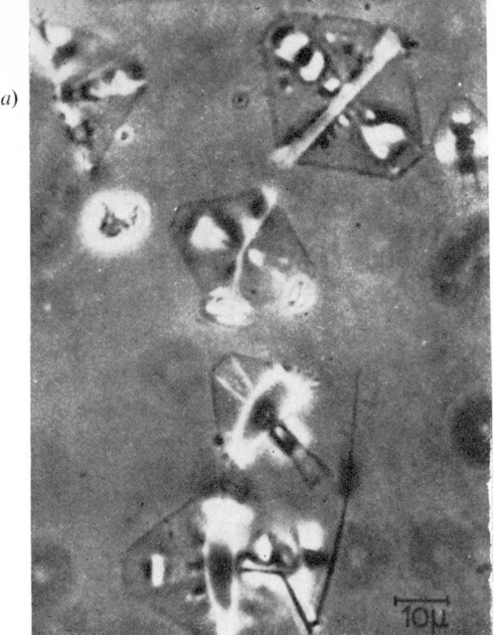

(b)

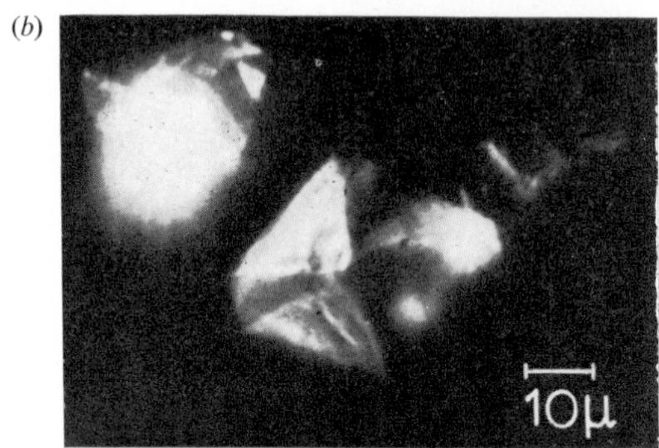

(c)

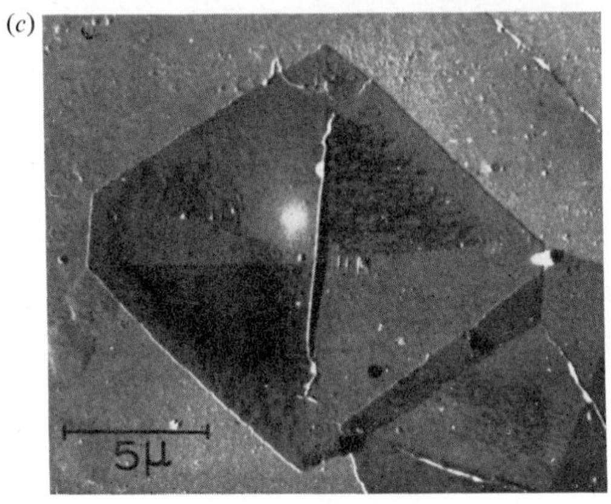

Plate II. (*a*) and (*b*): Micrographs of floating polythene crystals showing hollow pyramid structure. (*c*) Electron micrograph of a collapsed hollow pyramid (note central pleat). (From Keller 1964.)

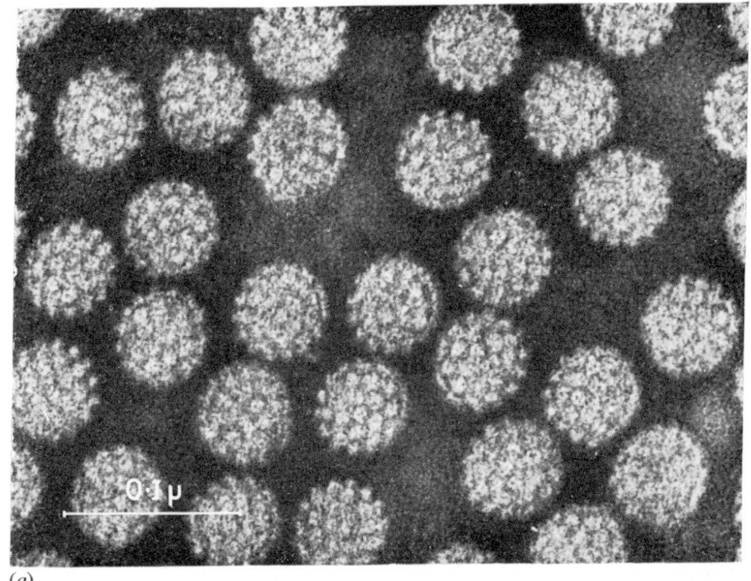

(*a*)

Plate III. Electron micrographs of virus particles. (*a*) (*above*) Human wart virus. Each particle has 72 protein subunits arranged on its surface. (*b*) (*below*) Adenovirus particles. Here 252 protein subunits form an icosohedral box of about 75 millionths of a millimetre across. (Particles negatively stained with potassium phosphotungstate.)

(*b*)

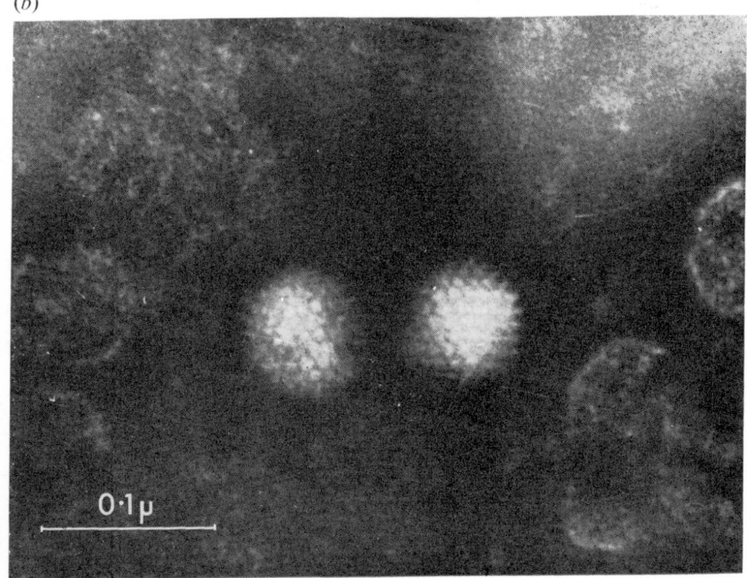

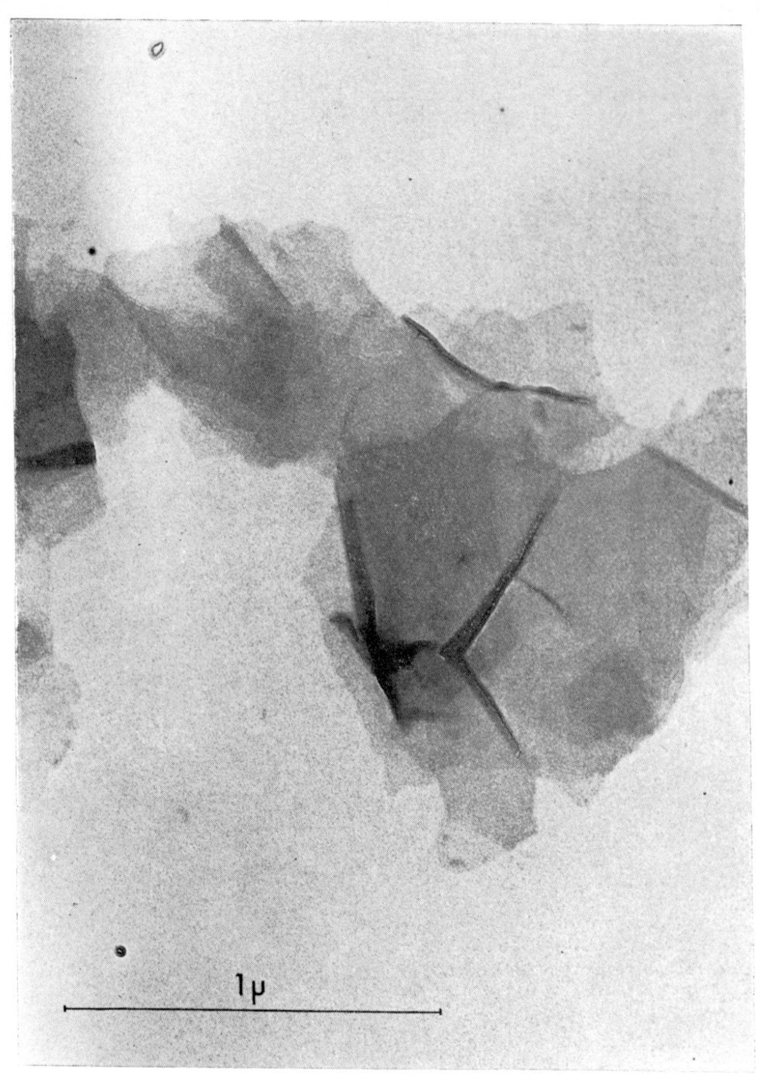

Plate IV. Electron micrograph of a montmorillonite.

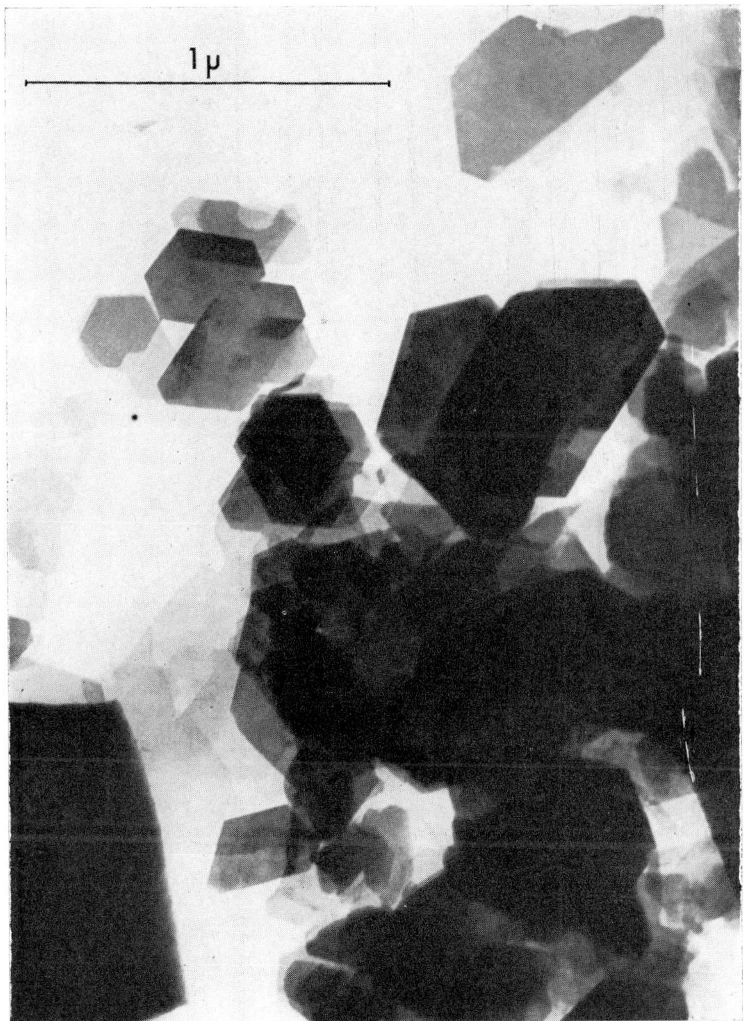

Plate V. Electron micrograph of a kaolinite.

Plate VI. Electron micrograph of a rectorite. (From Brown and Weir, 1963.)

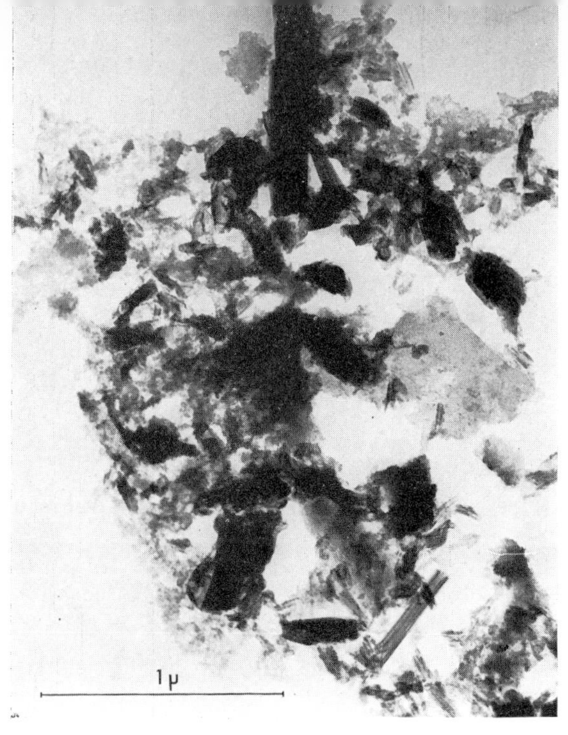

Plate VII (*left*). Electron micrograph of a halloysite.

Plate VIII (*below*). Electron micrograph of a synthetic Silicon-Aluminium-Magnesium clay. (From Caillère, Oberlin and Henin, 1954.)

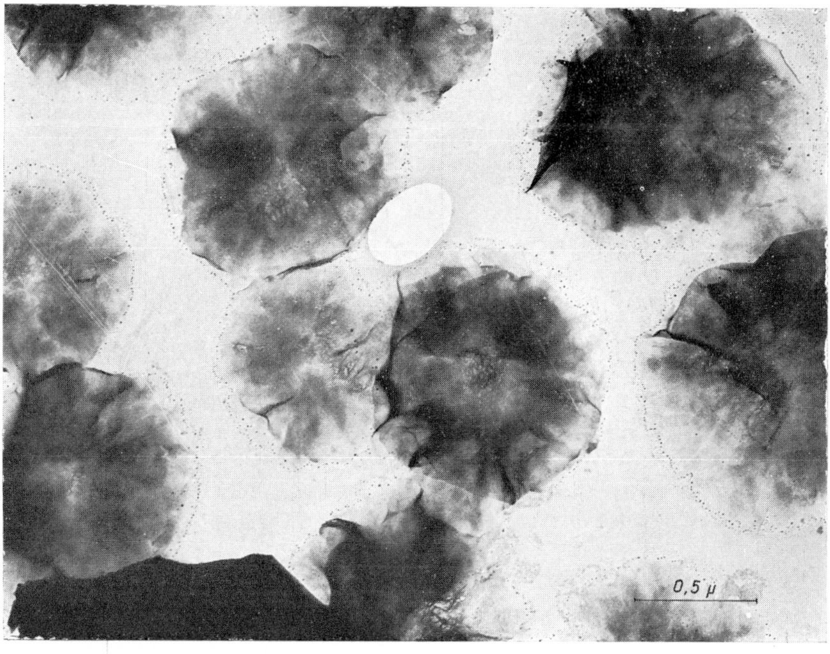

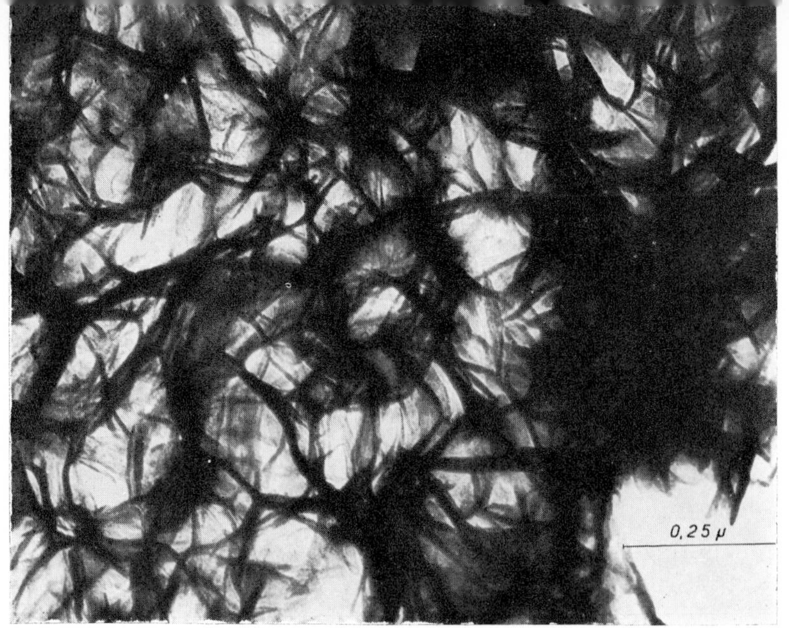

Plate IX (*above*). Electron micrograph of a synthetic silicon-magnesium clay formed at 20°C (pH 8) after several weeks. (From Caillère, Oberlin and Henin, 1954.)

Plate X (*below*). Electron micrograph of a silicon-magnesium clay formed at 100°C (pH 11·2) after four days.

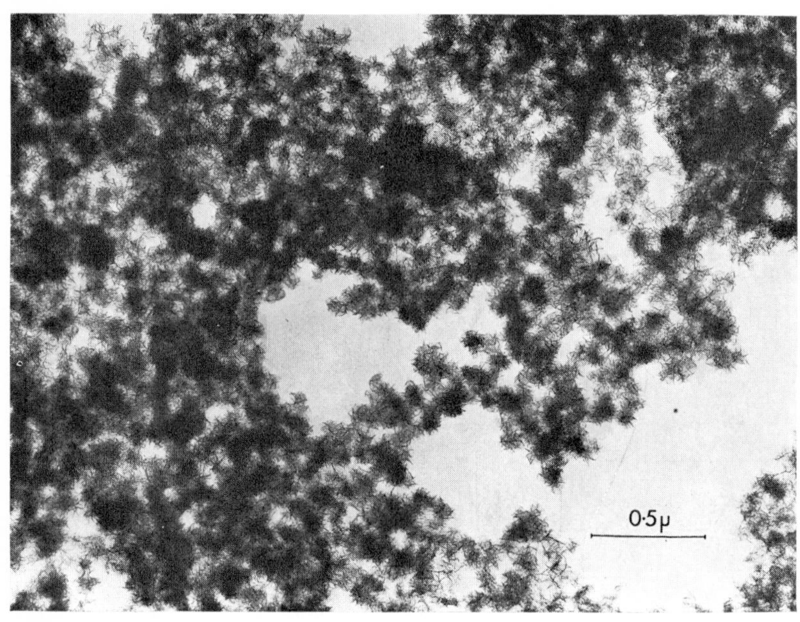

blind are at least very myopic. There are no means whereby, even figuratively, the primed RNA that we talked about could 'know' that by adding its amino acid to the growing protein chain it would be contributing to the general good of the huge molecular community of which it was a part. Letter-by-letter processes may have the advantage that they can organise a system very quickly (think of a freezing pond), but there is a price to pay for this slickness. The resulting organisation can only be local because the criteria of selection are local. If selection is to give rise to organisation that relates to a large whole system, to strongly co-operative organisation—the kind of organisation in the genes—then selection must operate on the whole system. But did not the monkey at his typewriter teach us that word-by-word selection was a slow business and sentence-by-sentence selection (never mind book-by-book selection) unthinkable? It would indeed be unthinkable as a required stage in the development of a phenotype, but fortunately there is a way round the problem for the evolution of genotypes. We might say that the trouble with whole-selection is that it implies whole-rejection. Hereditary organisations may be able to compensate for this loss by selective reprinting: they may then become increasingly efficient *as a whole* to a degree that is quite beyond non-replicative systems: hereditary organisations, and only hereditary organisations, can (sometimes) evolve. We will try to demonstrate this now through a series of 'thought experiments' that are marginally more realistic than the efforts of the would-be literary monkey.

HOW TO MAKE AN ENZYME WITHOUT REALLY KNOWING WHAT YOU ARE DOING

The problem we set ourselves is to design in principle a system that could automatically find and synthesise an enzyme from general functional specifications that we give it. Suppose you wanted to make an enzyme that would hydrolyse methyl acetate, i.e. break methyl acetate molecules into two pieces like this:

$$\mathrm{H{-}O{-}H} + \mathrm{H_3C{-}C({=}O){-}O{-}CH_3} \longrightarrow \mathrm{H_3C{-}C({=}O){-}O{-}H} + \mathrm{H_3C{-}O{-}H}$$

One approach in principle would be to think of a plausible transition state and then try to work out an amino acid sequence which would fold up in such a way as to create a groove that would stabilise this transition state (cf. chapter 3). This would not be easy. But in any case it is not the kind of approach that we want. With such an approach all the important organising would have gone on in your head.

Consider the general design shown in Fig. 27. The approach here is simply to synthesise amino acid sequences at random and then pick out those that happen to work. By taking apart the unsuccessful sequences and starting again, you will presumably eventually convert

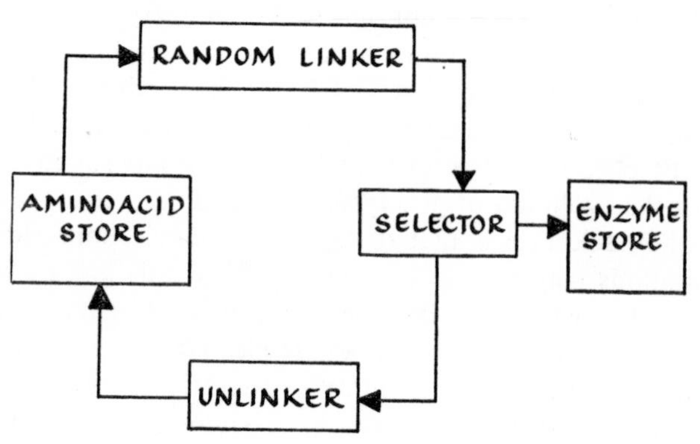

Fig. 27. First design for an automatic system for discovering enzymes.

all the amino acids into functional sequences—without knowing what they are or how they work. You would not expect any two of these sequences to be the same—not if they contain more than about a dozen amino acids—because it is wildly improbable that a long sequence would arise more than once within a reasonable time. As we remarked in chapter 3, there are 10^{195} different possible proteins of 150 amino acids—more than 'the number of electrons in the universe'. In practice you could ignore the possibility of hitting on one particular sequence of, say, 150 units. Only functions that could be carried out either by small sequences or by *a very large number of different sequences* could ever be generated by our machine. Suppose for simplicity that the random linker always 'deals' sequences of exactly 150 units, giving 10^{195} possible outcomes. If a million,

million, million, million, million of these possibilities are functional, i.e. 10^{30}, then this would still be far too few. There would only be $10^{30}/10^{195}$ chances, i.e. only one chance in 10^{165}, that the random linker would hit on a functional sequence on any one 'deal'. No, the chances must be very much better than this for the machine to work at a remotely practical rate. Something like at least one in a billion of the possible sequences must be functional—this would give, for example, 1 gramme of functional product for every million kilogrammes of material turned over. Of the total of 10^{195} possible sequences, then, something like 10^{186} would have to be functional.

Attempts to use chance and selection to put together very particular organised structures 'in one shot' in a reasonable time seem to be doomed whether one employs monkeys or automatic machines. Is there not some way, though, in which one might select amino acid sequences stagewise? Suppose that you started with a selector tuned to accept sequences with only a very faint activity for hydrolysing methyl acetate and then, having collected a batch of 'first improved' product, one puts the material through progressively more rigorous selection 'filters'.

RANDOM BATCH → select "good" sequences → FIRST IMPROVED BATCH

↓ select "better" sequences ↓

THIRD IMPROVED BATCH ← select "best" sequences ← SECOND IMPROVED BATCH

But selecting in this way, in stages, does no good since the successful sequences have to be produced in the first place—and the whole trouble is that if these sequences are very rare they will virtually never be produced. You might as well have done the final most rigorous selection on the original batch.

Suppose, then, that we introduce an element of shuffling into the system during the selection stages in order to create new possibilities. Suppose that after forming the first improved batch we introduce small random changes: not a complete reshuffle, just an occasional random change of single amino acids on the chains ('mutations'):

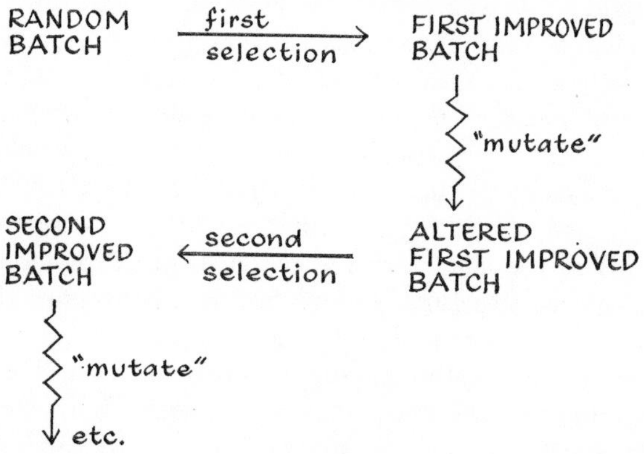

RANDOM BATCH —first selection→ FIRST IMPROVED BATCH

"mutate"

SECOND IMPROVED BATCH ←second selection— ALTERED FIRST IMPROVED BATCH

"mutate"

etc.

This fails again, because you are always more likely to disorganise than organise a system by any random change—however small. It would have been better to introduce the slight random alterations right at the start before any selecting had been done, because then at least there would have been a compensating chance that for every functional sequence that got spoilt, one of the much larger number of non-functional sequences would have become functional. It would have been better, that is, to have shuffled slightly an already completely shuffled system, that is not to have shuffled at all. This second stage-wise procedure* would be even less successful than the first.

HOW LUCKY IS INSULIN?

Let us now consider another 'thought experiment' of a highly idealised kind. We want to be as fair as we can to 'chance + selection' and see if it might manage, with really substantial resources and a super-efficient selection procedure to generate, as a complete unit, one particular protein of about the size of insulin. Let us say one particular sequence of just 50 amino acids.

---

*It seems to me that Oparin's suggested pre-reproductive evolution is essentially of this kind (Oparin 1957, pp 349–359.)

Once upon a time there was a planet of the mass of the Earth. There was no life on it, but various non-biological processes had resulted in the entire planet consisting of nothing but α-amino acids of the 20 kinds that exist in our proteins. Furthermore, all the amino acids were joined into random 50-long chains. As a result of a peculiar radiation from its nearby sun, mutations were occurring continually in the chains at an average rate of 1/amino acid/second. Each chain, then, was changing into a different chain about 50 times/second. Beside this flickering Earth-sized mass there sat an Incredibly Efficient Selector. The penetrating power of his vision, by which he could see any particular sequence he was looking for whenever it turned up, was only surpassed by the quickness of his hand with which he plucked out such sequences that took his fancy. In his other hand the Incredibly Efficient Selector held a bottle so that he could safely store the sequences he had chosen, away from the effects of the mutating radiation. How much of a particular 50-sequence would you expect him to collect in a period equal to the Earth's history (about five thousand million years)? The answer is about 40–50 molecules.*

The Selector would doubtless suffer from terrible feelings of frustration as he watched near misses come and go. Quite often he would see sequences that had only one unit wrong. There are 19 ways in which any given unit in the chain can be wrong and there are 50 units, so the total number of possible 'one-wrong' sequences is 50×19. 'One-wrong' sequences, then, should be almost 1000 times as common as correct sequences. Perhaps the Selector might be tempted to keep these near misses aside for special treatment: to mutate them 'carefully' perhaps. But unfortunately there would be just one way in which a single change in a 'one-wrong' sequence could result in the correct sequence, 18 ways in which the mistake would be changed for another one at the same site as the original mistake, and 49×19 ways in which a second mistake would be added.

---

* The mass of the Earth is about 6×10^{21} tons, about 6×10^{27} grammes. A 50-amino acid protein weighs about 10^{-20}g so the total number of 50-chains at any instant would be about 6×10^{47}. These chains change about 50 times/second and in 5×10^{9} years there are $1{\cdot}5 \times 10^{17}$ seconds so the total number of sequences tried out would be about $6 \times 10^{47} \times 50 \times 1{\cdot}5 \times 10^{17} = 4{\cdot}5 \times 10^{66}$. The total number of *possible* 50-sequences is 20^{50} about 10^{65}—so each sequence should turn up about 45 times during the period of the Earth's history.

So while it may be true that 'one-wrong' sequences are 50×19 times as common as correct sequences, their chances of maturing into a correct sequence at the next mutation are just one in $1+18+(49 \times 19) = 50 \times 19$. And this will be true whether the mutation is done ('carefully') on molecules that have been put into a special bottle, or on molecules left in the Earth-sized mass. As in our previous discussion a stagewise procedure makes no difference to the net outcome.

There is, however, a general way out. The reason for the failure of the stagewise procedure could be seen from the exact balance between the number of 'one-wrong' sequences which could be expected to appear by chance and their probability of maturing. To overcome this the Selector must cheat in some way. He must do more than simply select. He must upset the balance by recognising and actively helping the near-miss sequences. There are two possible approaches. He may prevent further changes in those parts of a sequence that are correct, or he may arrange for complete near-miss sequences to be copied as a whole many times whenever they appear. The first of these ideas is essentially a 'letter-by-letter' approach and could not give rise to co-operative organisation—only if whole sequences are on trial can you expect the resulting organisation to be a function of a whole sequence. It is the second idea that holds the secret to the consistent creation of highly co-operative organisation.

Suppose, then, that the Selector prints off a million copies of the first 'one-wrong' sequence to appear and then throws these back into the pool. The chances are that 1/50th of a second later he will have nearly 1000 correct sequences on his hands (i.e. 1 in every 1000 or so of the 'one-wrong' sequences will have matured). The Selector would not have to wait for a 'one-wrong' sequence to appear before starting selectively to amplify sequences in this way. He could start with any sequence that just happened to have at least one amino acid correctly placed. Such sequences would be very common accounting for over 90% of the material.* But even those sequences with as many as 25 amino acids in the correct positions would be common enough (about one in 10^{19}). At any instant the Selector would almost certainly be able to pick such a sequence out of a mass

* About 77% of the material would have either 1, 2 or 3 amino acids correctly placed, purely by chance.

of just a few grammes. He could then suitably 'amplify' his choice, throw back the copies, and then about 1/50th of a second later retrieve that fraction (about one in a hundred) which happened to mutate to give '26-right' sequences. He could then again replicate these—and so on. Even if the replicating process were quite slow, and took a whole fortnight every time, the Selector would still produce a correct 50-sequence within a year—a distinct improvement on the 50 million or so years which one would have expected to have to wait using the former extravagant, but non-replicative process.

But the trouble with a non-replicative selection procedure is not so much that it is *slow* in producing co-operative organisation. The trouble is mainly that it is in practice very *limited.* You might say (with some stretch of credulity) that a structure as specifically complicated as one praticular 50-amino acid sequence might have appeared on the primitive Earth: that if the origin of life had depended on some structure as particular and complicated as this then—even then—the origin of life would be reasonably explainable in physical terms. But such a structure is right at the edge of credibility. Had we wanted to generate a particular 60-sequence using our absurdly ideal non-replicative procedure, then we would have been in for a long wait indeed. The expected period would have been about one hundred thousand million *times* the age of the Earth. By contrast, our ideal replicative technique would not have taken much longer to generate a 60-sequence than a 50-sequence.

We may draw the following interim conclusions. Blind chance is a creative fellow. In suitable circumstances he can produce any *kind* of organised arrangement, in particular co-operatively organised arrangements. He is, however, very limited. Low levels of co-operativeness he can produce exceedingly easily (the equivalent of letters and small words), but he becomes very quickly incompetent as the amount of organisation increases. Very soon indeed long waiting periods and massive material resources become irrelevant.

Beyond setting the stage and providing components, it is difficult to see the point of a protracted pre-reproductive 'evolution' which a number of writers invoke to generate the first fully reproducing systems. Such an 'evolution' could not have produced any kind of co-operativeness that Blind Chance could not have produced. *Only* hereditary organisations are capable of improving significantly on Blind Chance in this crucial respect. The *first* hereditary organisations, then, could not have been more than trivially co-operative. This immediately rules out, for example, any system using *both* nucleic acid and protein in anything remotely resembling their modern roles. I believe it also rules out any system using *either* nucleic acid or protein, but we will leave further discussion of this until chapters 7 and 8. We will now try to clarify the conditions under which hereditary organisations can evolve.

CHANCE, SELECTION AND REPLICATION

It may reasonably be objected that the above 'discovery' of a particular amino acid sequence depends on a Selector who knows the answer already. Indeed he knows it so well that he can recognise sequences that are 'on a route' to the sequence that is being sought. This objection does not apply, however, to functional as opposed to structural selection procedures. In attempting to discover an enzyme by an automatic selection procedure (Fig. 27) we were not concerned with arriving at a particular structure, but at a particular ability—the ability to hydrolyse methyl acetate.

Now consider the design outlined in Fig. 28. To begin with, random sequences are fed into a selector as before. The selector must be tuned to let through even quite feebly active sequences—because this is likely to be all it will be able to find. Suppose it lets through the best 1% of the random sequences. These successful sequences are then used to direct copies of themselves (×100) in the Replicator. Separate amino acid units are taken from the store for this purpose. A mutator then produces a small number of random changes affecting only one or two amino acids in only a proportion of the sequences. The material then starts to re-enter the selector. We can consider these new candidates as falling into three categories. There would be sequences which had been unaffected by mutations: these would clearly be passed by the selector because their identical parents had been passed. Then there would be a fair number of sequences which had been spoilt by mutations and which would probably fail this time round. Occasionally there would be sequences that had been improved by a mutation: these would certainly pass. As soon as the selector started to deal with these second-generation sequences the flow meter would register a substantial increase in number of successful sequences. The selector could then be re-tuned automatically by a signal from the flow meter so as to maintain, say, a 1% pass rate.

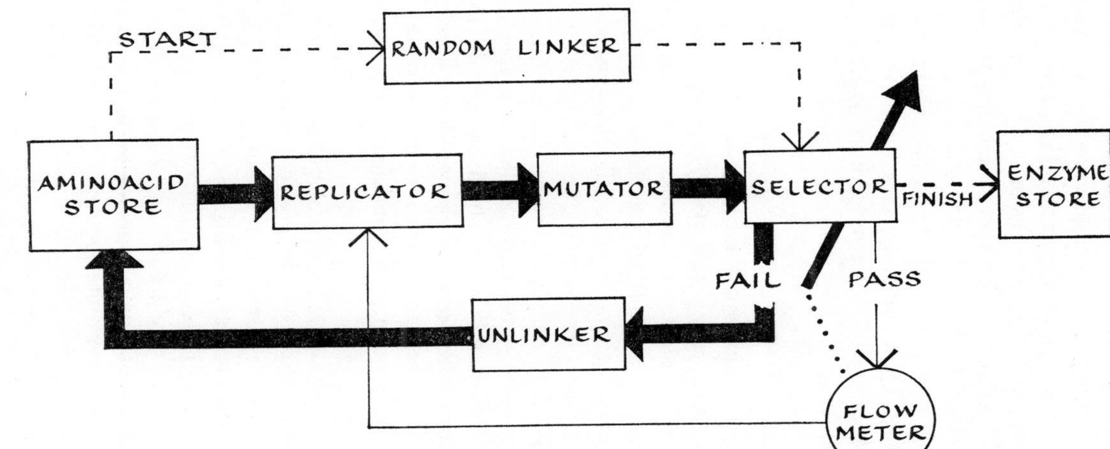

Fig. 28. Second design for an automatic system for discovering enzymes (cf. Fig. 27).

In this way those rare sequences that had been improved by a mutation would be strongly favoured. They would be almost certain to have offspring. The rare successful mutations in these offspring would again be selectively amplified—and so on. In such a way increasingly efficient sequences could apparently be formed without anybody having any idea of their structures or how they worked. You would see the pointer on the variable selector moving progressively towards higher and higher standards. When it reached some desired level, the material going round the cycle could be automatically shunted into an enzyme store. Quite suddenly, then, almost all the contents of the amino acid store would be converted into highly active sequences in the enzyme store.

CORRESPONDING CONTINUITY OF FORM AND FUNCTION

The kind of automatic evolution which we have been discussing could only work if there were a reasonably strong tendency for similar sequences to have similar functions. It is necessary that a small random change in a moderately efficient sequence should have a better chance of yielding a very efficient sequence than a small random change in an inefficient sequence. This is an obvious enough idea, but it is of great importance for the evolution of organisms. It is an example of a general and familiar principle that there tend to exist correspondences between small changes in form and small changes in function. A random change in an uncomfortable chair may not be very likely to turn it into a comfortable one, but it is *more* likely to do so than a random change in a *very* uncomfortable chair.

There are usually quite broad corresponding continuities of form and function for objects such as pieces of machinery or sculpture. Let us say that such things show a 'strong CCFF'. Languages, on the other hand, do not show a strong CCFF. A random change in a word may possibly produce another word with a meaning, but it will not be likely to produce another word with a similar meaning. This is a weakness of language analogies for proteins—because proteins probably do have quite a strong CCFF. It certainly looks as if this will be broadly true for the tertiary structure of proteins, but more importantly from our point of view it looks as if it is true also for primary structures. We know of many examples of similar protein sequences with similar functions, and it seems likely that many, if not most, single arbitrary changes of an amino acid in a protein chain produce only minor changes in tertiary structure and hence (usually) only minor changes in function.*

ACCESSIBLE AND INACCESSIBLE PROTEINS

To see more clearly the importance of the CCFF idea in relation to the evolution of proteins we may consider a point, raised by Maynard Smith (1962), that there might exist groups of proteins that are inaccessible to evolutionary processes. Let us consider a particular imaginary example. Suppose that one way of making a protein that could hydrolyse methyl acetate is to join together 150 amino acids with the first 50 in some specific order and the other 100 in any order at all. Suppose further that for the first 50 amino acids there is no continuity with similar sequences—that any mistake here completely destroys the function. Although there would be only one way of arranging the 'fussy' part of the molecule there would still be 20^{100} (about 10^{130}) ways of arranging the 'unfussy' part. We could say that out of all possible sequences of 150 amino acids that can hydrolyse methyl acetate there exists a particular set of about 10^{130} different proteins any one of which would be functional. Yet this set would be too *small* a sub-set to be hit on by chance—it constitutes one part of 20^{50} (i.e. about one part in 10^{65}) of the total number of possible 150-sequences. And it could not be evolved towards either because there are no approach routes to the 'fussy' part of the molecule.

What kind of situation is required, then, for a functional protein to be evolutionarily accessible?

Suppose that you could represent the set of all possible 150 amino acid sequences by the area A (Fig. 29). We imagine that each possible sequence is represented by a particular tiny area within A and that these tiny areas are so disposed that structurally related sequences are close together and unrelated ones far apart.† B, C and D might

* Some single amino acid changes, e.g. at an active centre, may be disastrous. But we can put up with such things. What is important is that many changes should give rise to only small alterations of function. The subcrystalline structure of enzymes (cf. chapter 3) provides a situation in which the conformation of the active centre depends indirectly on very numerous units, some quite distant. This looks like the required kind of situation, with great scope for 'fine adjustments' through changes in amino acids that are far away from the active centre (p. 47).

† It is only fair to say that this is impossible: one would need a multi-dimensional space as for the epigenetic landscape. But again we take the liberty of representing a large number of dimensions (here 150) by just two.

represent different sets of sequences able to hydrolyse methyl acetate through radically different approaches. Members of any given set are structurally related to each other in a continuous way but unrelated to members of other sets. They form isolated islands. Such sets of sequences will be accessible only if they are big enough to be hit on by chance. The particular inaccessible set which we discussed,

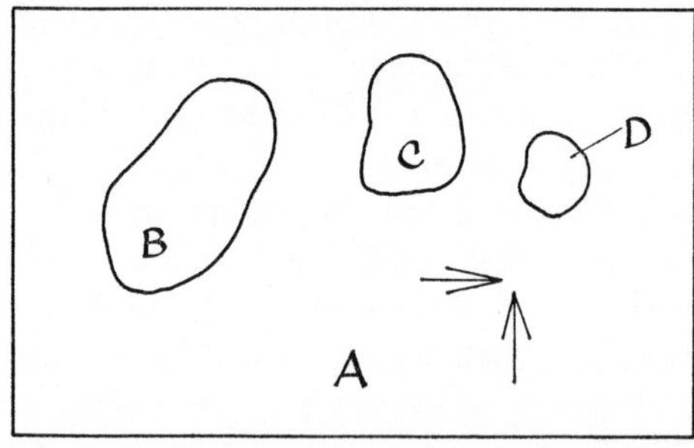

Fig. 29. Functional sub-sets of sequences of a given length are represented by zones (B, C, D) within the larger area (A representing all possible sequences of that length).

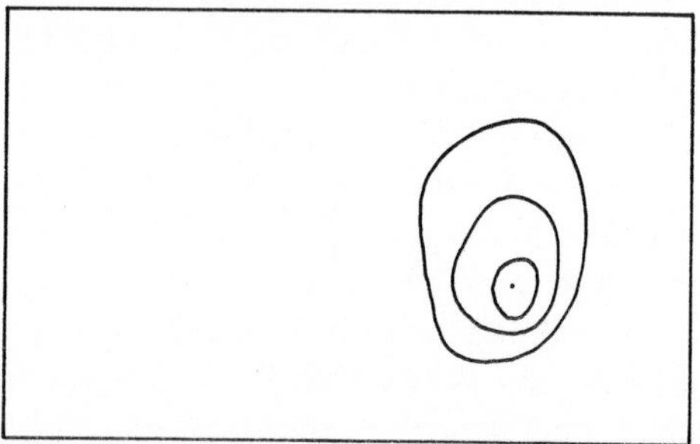

Fig. 30. Small efficient sub-sets may be 'accessible' if they lie within less efficient but larger sub-sets (cf. Fig. 29).

containing the 'fussy' 50-sequence, is indicated on the diagram by the pair of arrows. It is so small that the diagram would have to be magnified a billion billion billion billion billion times before it would become visible. Such an exceedingly small set could only become accessible if it lay within a hierarchy of larger and less efficient sets as in Fig. 30. It does not matter how small the most efficient set is, so long as the largest set to which it is related in this way is large enough to be hit on by chance.

Our Incredibly Efficient Selector made use of such an idea. He started with a set of sequences—'25-right' sequences—which were common enough to appear by chance. He then proceeded to the smaller '26-right' subset and finally arrived at the 'remote' 'all-50-right subset. The required continuity of form and function was guaranteed by the artificial device of defining efficiency structurally: the sequence was taken to be 'better' simply if it has more of its amino acid corresponding to some arbitrarily chosen ideal sequence. Such an 'evolution' could be represented on a rather simple kind of 'target' diagram—see Fig. 31. A shot landing anywhere on the target will 'migrate' through selective replication of random mutations until it arrives at the bull's eye (see Fig. 32). An alternative way of looking at it is to regard the circles as contours defining a conical depression in the ground—like the 'brewer's 18th' (p. 82). The 'swarm' then behaves like a ball placed somewhere in such a depression. It rolls down to the most efficient sequence.* We will use these alternative views interchangeably.

Where selection depends on a functional criterion that is not so rigidly related to amino acid sequence, the corresponding diagram will be more complicated. It will be some kind of landscape such as that indicated in Fig. 33. There will be many 'bull's eyes' corresponding to different structural approaches to the function. Radically different approaches may belong to quite separate zones: different approaches in detail will correspond to multiple 'bull's eyes' within a given 'target'. And the targets will be highly distorted. This corresponds to the idea that the relationship between the primary structure of a protein and its functional efficiency is not uniform in all possible directions: some amino acid changes will have more effect than others.

Steep cliffs would be a feature of some landscapes—as in zone (*a*) Fig. 33. These would correspond to some particular approaches depending critically on some particular amino acids or groups of amino acids: *some* directions of change in structure do not produce a continuous change in function (see footnote p. 99).

---

* Note that we are regarding downhill as more efficient. It is more usual to regard evolutionary processes as seeking a hilltop rather than the bottom of a basin (Pask, 1961). The arguments are not, of course, affected: but a downhill model is perhaps easier to think about here as it is in line with other models which we have used of balls rolling on surfaces seeking low points.

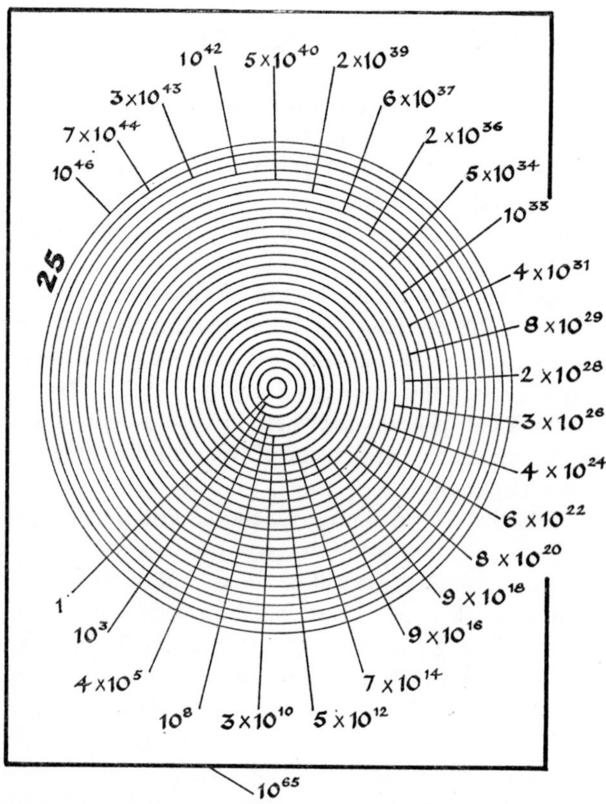

Fig. 31. The technique which the Incredibly Efficient Selector used to 'evolve' a particular 50-amino acid sequence can be discussed in terms of a 'target'. The total area within the square represents the set of all possible 50-sequences (there are about 10^{65} members of this set). The total area within the largest of the circles represents the set of sequences with at least 25 amino acids correctly placed. The areas enclosed by the smaller circles represent the set of sequences which have at least 26, 27 . . . 49, 50 amino acids in the correct positions. The numbers give the approximate membership of each of these sets. These numbers are not remotely proportional to the areas included by the circles. To be in scale such a diagram would have to be multi-dimensional and of cosmic capacity.

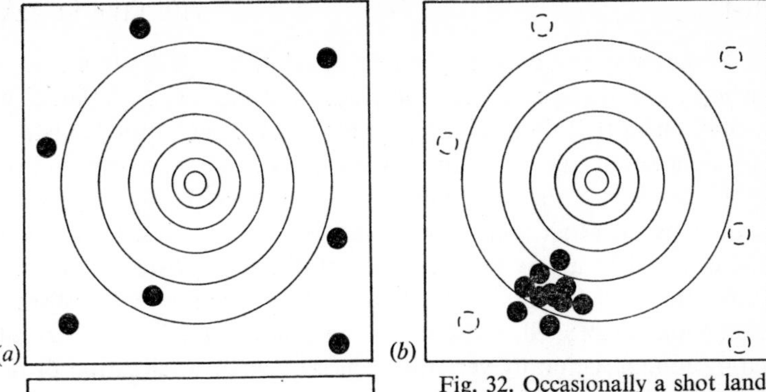

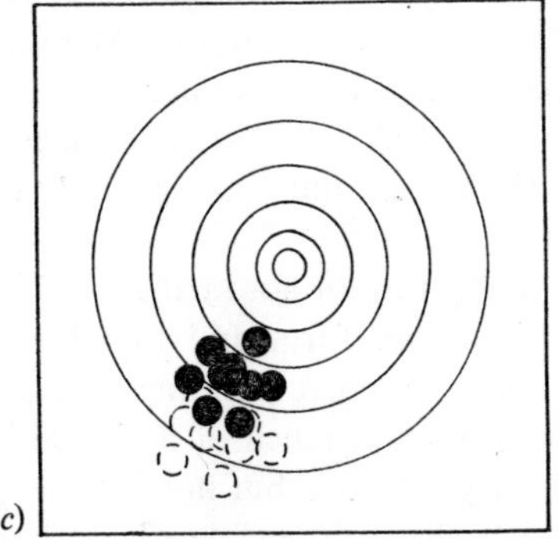

Fig. 32. Occasionally a shot lands on target (*a*). This is selectively replicated with occasional small changes giving rise to a local 'swarm' (*b*). A few members of this 'swarm' happen to spill into the adjacent smaller but more efficient zone and create a new 'swarm' through selective replication (*c*). The overall effect of selectively replicating the more efficient sequences is that the 'swarm' is inevitably drawn towards the central most efficient sequence (cf. Fig. 31).

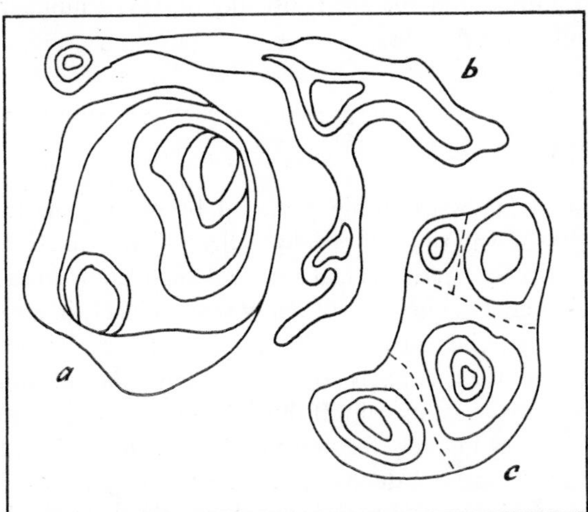

Fig. 33. The kind of 'landscape' that might be explored by an automatic enzyme-finding machine. Some zones (like *a*) might have steep cliffs. Some (like *b*) might include very different sequences. In general you would expect many local minima in each zone: (*c*) has four, each with its 'catchment area' indicated by the broken lines.

Two amino acid sequences might be related to each other and yet have few if any amino acid positions in common. By 'related' we would mean that they belonged to the same zone on the landscape, e.g. zone (*b*) in Fig. 33. In other words, it would be possible in principle to go from the one sequence to the other through a series of unit steps such that each intermediate is itself functional. In general there would be a very large number of different particular routes—particular walks across the landscape—by which such a transformation could be effected. But some of the zones might be quite straggly connecting structurally very different sequences. A situation of this kind appears to exist in the haemoglobins and myoglobins. Members of this group from different sources have closely similar functions and tertiary structures, but often very largely different sequences (Zuckerkandl, 1965). (See Neurath, Walsh and Winter, 1967, for other examples—among enzymes.)

To say that the evolution of a protein in an automatic machine requires a reasonably strong continuity of form and function in proteins is equivalent to saying that the corresponding abstract landscapes should be good walking country: much of it should be rolling landscape. There can be some precipices, but not too many. The kind of landscape that would be least amenable would be a completely discontinuous one consisting of, say, a number of mine-shafts drilled in a flat plain.

WOULD THE AUTOMATIC ENZYME FINDER ALWAYS COME UP WITH THE SAME ANSWER?

The abstract landscape being explored by an automatic machine of the kind shown in Fig. 28 would be fixed by the function that was being sought. We should remember that such landscapes for fairly long amino acid sequences are really vast—bigger, in a sense, than the known material universe. The number of local depressions in the landscape, then, is likely to exceed the number of molecules in the machine by a vast margin. If this is so in any particular case then it would be very unlikely for two randomly generated sequences to land within the same 'catchment area'. So to begin with, during an evolution within our automatic machine, there would be a very large number of individuals giving rise to separate species evolving towards different goals, i.e. towards the lowest point in their own private local depression. Those in very shallow basins would soon go extinct, thus releasing material to swell the numbers of the more fortunate ones in deeper depressions. These could continue for longer in competition. If the standard required were not too high, there might thus be a very large number of different species of molecules in the material eventually shunted into the enzyme store. But, in principle, provided the Selector was told to find a 'best possible' sequence, you could end up with only one kind of sequence in the store—a pure protein. This would correspond to the lowest point within the landscape *which had been explored.* (The lowest known point, so to speak.) But if you started all over again the initial random sequences would land on a different collection of catchment areas. Again there would be a single best sequence corresponding to the lowest point in *this* set of catchment areas. Again the machine would produce a pure protein if you ran it long enough and continued to press for as high a standard as possible. But this time (I would guess) it would be a quite different protein that would emerge as 'the best possible'.

REAL EVOLUTION OF PROTEINS AND ORGANISMS

Any particular protein within an organism is subject to similar forces to those that would operate within our automatic machine. Proteins are never, so far as we know, replicated; but they tend to be selectively reproduced if they are efficient components of organisms. Broadly, a protein sequence is 'efficient' if it contributes to the likelihood of the survival for reproduction of the total 'community' of protein sequences to which it belongs. The members of this 'community' have many various ways of contributing. The overall function—to leave progeny—results from an interplay of subfunctions. Each subfunction has its own 'landscape'. The evolution of an organism is a simultaneous exploration of as many different landscapes as there are functionally distinct proteins within the organism. And many of these subfunctions that are thus actively selected for can be exceedingly subtle. It may be very important for an organism to be able to carry out a particular reaction only under quite narrowly defined circumstances. It may be a matter of life or death whether a membrane protein is so constructed that it will allow, say, sodium but not potassium ions to pass. The subfunctions in question may be seemingly obscure and remote from the rather simple overall function of the organism: as obscure and remote as a particular screw in a car engine might seem to someone who can see what a car is for, but has never understood the point of a carburettor.

But why are organisms so complicated? Organisms are not only highly unified in the sense that their overall function is critically dependent on a collaboration of subfunctions, but in many organisms, man for example, the complexity of interplay of subfunctions seems to have been taken to quite unnecessary lengths. Unnecessary, that is, in terms of the overall function—ability to leave progeny. Why did organisms bother to evolve beyond bacteria? Bacteria are very good at leaving progeny. It is as if, having settled on a good general way of coping with the primary biological function aeons ago—having created the nucleic acid-protein biochemical system—evolution has since been mainly concerned with almost baroque elaboration. (To change the analogy, one is reminded of the hi-fi enthusiast who said that you should never use two loudspeakers when six will do equally well.) By itself natural selection does not explain the cow. A cow, with all its remarkable subfunctional parts (horns, tail, eyes, udder and so on), is still no more biologically efficient than a bacterium.

In the idealised evolution of a protein in our machine, we expected that the system would move from a state in which many different species were competing to a state in which only a few—finally only one—were left. This is not what seems to have happened in nature to organisms. Rather the reverse: our common biochemistry indicates a single ancestor for all the profusion of modern life forms. Evolution has been a divergent branching process.*

A third related problem is that many organisms seem to have gone to great lengths to survive and thrive in the most unpromising environments.

We might say, then, that while natural selection can in principle explain the evolution of unified and biologically efficient (progeny-leaving) organisations, and while it may explain the appearance of complex systems in so far as the complexity contributes to efficiency, it does not in itself explain the 'unnecessary' complexity, nor the diversity nor the ubiquity of known living forms. Can we improve our 'landscape' models to make them more closely fit the facts?

---

* This aspect of evolution has been described by Rensch as cladogenesis, as opposed to anagenesis leading to more efficient organisation. (See Huxley, 1963, p. xxxii).

MOBILE LANDSCAPES AND SUPERLANDSCAPES

In the enzyme finding machine the abstract landscape was fixed by the fixed criterion of efficiency. In an organism, however, the very various criteria of efficiency for the different proteins are interdependent. As one protein evolves on its landscape it may alter the landscapes of other proteins by altering functional requirements. One should think of the sum of all the protein sequences as a single unit—one should think in effect of the total genotype of the organism as part of an even more fabulously vast landscape of even higher dimensionality—the landscape of possible DNA sequences. The evolution of a given species, then, will consist of a single journey (of a 'swarm') on this superlandscape.

Now such a superlandscape will not be a static affair either. Thus even if a genotype reaches a local minimum it will not necessarily be stuck there for ever: it may be able to move across to an adjacent basin through more or less temporary 'earth movements' (compare Waddington, 1957, p. 112). These earth movements will result simply from changes in the real environment of the organism, for example, a change in climate or the appearance of competitors or predators. Or it might result from migration. If a group of organisms of a given species changes its environment by moving across the real geographical landscape this would correspond, not to a movement *across* the diagramatic landscape of DNA sequences, but, in the first place, to a movement *of* the landscape of DNA sequences. In so far as this may now alter the position of the local minimum in which the organisms were originally established, the geographical displacement may lead to the displacement of the group on the landscape of DNA sequences.

But it is still not clear why such antics should tend to lead to more complicated organisms *per se*. One could argue indeed that it is the most important question that remains. It is more or less what we mean by (long term) evolution that organisms have on the whole become more *interesting*. They have not only elaborated fascinatingly complicated, strongly interdependent networks of subfunctions, but they have invented all kinds of new ones, like walking and thinking and talking. It is the subfunctions that make life worth living.

Consider the simpler question: 'Why does a gas expand?' It is not because of any specific driving force in the sense of a repulsion between the particles or a 'desire' to go anywhere in particular. It is simply that a disorganising factor (here random thermal motion) will tend to move a system from a state in which there are relatively few possibilities into a state in which there are more. Even if the number of possibilities in the first state is Archimedian, if the number in the second state is much more Archimedian then that is the way in which the system is virtually certain to go. The effectively random 'earth movements' in the landscape of DNA sequences does not only push genotypes about: it creates a situation in which they will tend to move 'outwards'—become longer and more complicated—because there are more particular ways to go in that general way.

DYNAMICS OF A TORTOISE–BLANKET–PEA SYSTEM

Think of a very large blanket covering a dense crowd of randomly browsing tortoises. A handful of peas placed in the middle of the blanket would first settle into a local depression and then probably become scattered into progressively more numerous separate clutches, in separate local depressions. Eventually all the peas would roll off the edges of the blanket. One's overall impression would be of a divergent process—like a gas expanding. But looked at over a short period of time one might notice rather the continually convergent influence of gravity: this is the immediate force operating on the peas and its immediate effect is to move the peas in given local regions ('catchment areas') towards optimal (lowest) points. The process would be locally convergent and globally divergent.

A less fanciful model is a puff of steam: here again there is a convergent and a divergent process superimposed. The water molecules condense into droplets while the droplets diffuse away from each other. The universe itself is another model: the galaxies condense out of intergalactic matter while at the same time they are moving away from each other.

Like so many processes, then, evolution is neither simply convergent (it has no ultimate 'goal') nor simply divergent (it *has* local 'goals'). Evolution is driven by two opposing factors. One is analogous to the energy effect in physical systems. This is natural selection. The other is analogous to the entropy effect. This is environmental variability. Natural selection we see as a 'local', organising, factor: it is represented as movement *across* the landscape of DNA sequences: it seeks local sub-sets of greater biological efficiency while maintaining unity of organisation. Environmental variability is a 'global', disorganising, factor: it is represented on the model as a movement *of* the landscape: it tends to lead to the exploration of wider fields of possibility, and hence eventually to the characteristic variety and complexity of evolved forms.

FURTHER READING

WADDINGTON, C. H. (Ed.) 1968, 1969, 1970. *Towards a theoretical biology: 1 Prolegomena, 2 Sketches* and *3 Drafts*. Edinburgh University Press, Edinburgh and Aldine Press, Chicago. These are concerned with attempts to formulate general concepts characteristic of living systems. It may be, as Waddington says, that there is yet no clear subject 'Theoretical Biology', but there are theoretical biologists—this series gives a good idea of the ways in which they think. See also WADDINGTON, C. H. 1957. *The strategy of the genes*, Allen and Unwin, London.

PASK, G. 1961. *An approach to cynbernetics*. Hutchinson, London, is an elementary account of mechanical self-organising systems.

JOHNSON, H. A. 1970. Information theory in biology after 18 years. *Science* **168,** 1545. This is a non-mathematical discussion of concepts.

PART THREE—THE ORIGIN OF LIFE

7

A Quest for an Easy Life

You may regard the origin of life as a part of history, as a puzzle to be solved by historical detective work. You may perhaps examine the most ancient rocks available for evidence of life (see Eglinton and Calvin, 1967): or you may adopt a more indirect approach by first attempting to deduce, from geological and astronomical data, the general conditions that prevailed on the primitive earth (see for example Oparin, 1957; Firsoff, 1967; Sillen, 1967; Cloud, 1968) and then partly re-enacting the scene through experiments like the famous one of Miller (1953) to see which particular molecules may have existed in primitive waters. Such experiments indicate that the very molecules that are fundamental units of all present day terrestrial life might well have been present on the primitive Earth as plausible products of plausible processes induced by various available forms of energy—such as solar radiation and lightning (Calvin, 1965, 1969; Fox, 1965).

For reasons that I will try to make clearer later, I am inclined to think that these experimental results may be misleading—at least when applied to the origin of life itself, as opposed to its early evolution. For the puzzle of the origin of life there exists an alternative special approach that is much less historical: it uses our knowledge of chemistry together with the most general theories of biology. The approach is this:

(i) Take a good look at atoms and see what they can do,

(ii) Take a good look at organisms and see how they work, and then

(iii) Put an organism together in the easiest way you can think of. Do not be concerned if your design bears no chemical resemblance to the modern wonder-machine. Only two general principles should guide your first design:

(a) The system should have an indefinite potential for evolution, and

(b) It should be easy, technically, to make. (It may, of course be very difficult to think of.)

'THE FACILITY PRINCIPLE'

I shall make considerable use of this idea—that making an indefinitely evolvable system of some sort should be technically *easy* rather than merely theoretically feasible. This cannot at the moment be proved but the following general considerations make it probable.

One can speculate about the molecules and larger structures that may have been present on the primitive Earth. But it is instructive to think for a moment about things that we can be virtually certain were *not* there. Tin-openers for example. (Can you see a tin-opener washed up on a primordial beach?)

By using our intelligence, we can quickly create a functional assemblage of parts which belongs to a sub-set of possibilities that is so small that we can effectively rule out the thought of the spontaneous formation of that assemblage through a happy coincidence of unguided events. As we discussed in chapter 6, Blind Chance is a creative but very limited fellow. Admittedly he works best at the molecular level (it is perhaps unfair to expect him to produce a tin-opener), but even at the molecular level we can outdo him in putting together specifically, functionally complex arrangements. With the most ideal conditions and the most able assistant, Blind Chance was quite unable to shuffle together a specific protein as big as an insulin molecule in a respectable time (pp. 92-94). Using the latest techniques of chemical synthesis, we can put together billions upon billions of such molecules in a few weeks. Knowing the structure of, say, insulin, we can synthesise it readily enough. Now if indeed Blind Chance, under less than ideal conditions with only imperfect selection mechanisms, *did* manage within a few billion years to shuffle together some system capable of initiating the process of organic evolution, then this system, whatever it was, could not have been all that difficult to make. I would guess that we should manage a similar, if not identical, feat of initiation within a few days—*if we know what to do*. Conversely if we have so far failed, then it is perhaps our theories rather than our experiments that are at fault—they have lead us to consider the wrong kinds of system.

THREE IDEAS

Oparin (1924) and Haldane (1929) are generally regarded as the initiators of modern thinking on the origin of life. Since then, many other prominent men of science have devoted themselves to this subject. It has become thoroughly respectable, and there is now a very considerable literature. I refer readers to the bibliography at the end of this chapter for general accounts of ideas and experiments relating to the origin of life. Here I will simply abstract three key ideas that have emerged. One can very roughly categorise different views on the subject according to which of these ideas are included or are prominent in them.

The Idea of Original Biochemical Similarity

One of the most striking aspects of present day terrestrial life is the fundamental biochemical similarity of all its forms. Every organism, from that patch of dry-rot in the garage to the man who came along to take it away, seems to use nucleic acid to hold and print genetic instructions and protein to carry them out. This general observation, together with the strong implication that most of the fundamental units (amino acids and so on) would have been formed through non-biological processes, has lead to the very prevalent idea that the first organisms on Earth had a similar composition to modern forms, even if their constitutions may have been very much simpler. According to this idea, the first organism arose through some kind of association between suitably pre-formed parts which were similar to those out of which present day organisms are made. Indeed it is commonly held that these molecular parts are essential for life—or at least for a life possible or relevant under terrestrial conditions. A notable dissenter from this simple and rather dogmatic conclusion is Pirie (1959) who has argued that the modern biochemical uniformity is a product of evolution—a tribute to the effectiveness of natural selection—rather than an indication that life depends uniquely on amino acids, purines and so on. According to Pirie, life started in many diverse ways and then stabilised on the most efficient (see Marquand, 1968).

The Idea of Pre-reproductive Metabolism

Oparin has always taken a 'metabolic' view of life and its origins. According to Oparin the most essential aspect of life is dynamic. Life is 'a particular, very complicated form of the motion of matter' (Oparin, 1957, p. xii). Life evolved first through the establishment of separate non-reproducing open systems—gummy droplets in the sea—within which complex integrated networks of self-maintaining reactions had become established. Such systems were already subject to a simple selection in that those droplets in which the reactions lead to greatest stability would survive for longest, and would thus be most prevalent. At some stage, according to Oparin's view, some of these dynamic systems would acquire the ability to reproduce. They would then be subject to natural selection—Darwinian evolution would be under way. A similar position has been taken by Bernal, except that in his view the pre-reproductive 'metabolic' processes were not confined to separated systems to begin with, but spread throughout 'subvital regions' containing organic molecules held in proximity to each other on mineral surfaces—particularly on clays. According to Bernal, fully reproducing systems only appeared with the (comparatively late) appearance of nucleic acids.

The Idea of an Original Gene

Haldane (e.g. 1954) together with a number of other early thinkers on the subject (see Oparin, 1957, pp. 95–99), took a more genetic view of the origin of life. According to this view, some kind of accurately self-replicating molecule would have been an essential component of the very first organisms. The most extreme proponents see our ultimate ancestor as having been a 'naked gene'—a self-replicating molecule able to mutate randomly and pass on such changes to its progeny—which were, of course, other self-replicating molecules. Protein was originally the favoured material in such hypotheses (Blum, 1955), but since the discovery that DNA is the genetic material of present day organisms, some more or less 'clothed' nucleic acid is generally taken, by the genetically minded, to have constituted the first living system on Earth. This kind of view seems to be particularly acceptable to evolutionary biologists such as Simpson (1950), Huxley (1963) and Maynard Smith (1966). It has been very clearly stated by Horowitz (1959).

DIFFICULTIES PRESENTED BY 'METABOLIC' AND 'GENETIC' VIEWS

The 'metabolic' view suffers from difficulties that we discussed at length in the last chapter. We may agree that very complicated chemical reactions would have taken place on the primitive Earth. Indeed one can envisage the formation of gummy droplets in the sea, and of great volumes of adsorbed organic molecules on clays; and one can see these processes being continually fed through the production of reactive molecular species in the atmosphere as the result of ultraviolet radiation—on so on. But whether such processes could ever have become highly integrated in the kind of way that Oparin suggests; whether they could ever be remotely regarded as metabolic; whether, before acquiring the ability to reproduce, the chance acquisition by a system of some useful facility would really be of any help for the future in the absence of a mechanism for passing on the good news to off-spring; whether indeed such a system, even if it could become highly organised in a co-operative way, would show any tendency to acquire the ability to reproduce—all these points are open to question. From the discussions in chapter 6 it would seem that any organisation that depends at all strongly on the co-operation of parts can only arise in nature through the selection of whole co-operative systems following trials of relatively limited parts or aspects of the system—and this selection of wholes demands reproduction of wholes if it is to generate more than a trivial degree of integration of parts. In forming an apparently purposeful co-operative system, Blind Chance can manage to produce such things as crystal seeds, but this is about his limit. And waiting for a long time, or using massive material resources, is not, as we saw, much help.

Now if evolution depends on reproduction then it depends also on replication as an essential part of reproduction (see chapter 4): evolution depends on the pre-existence of something like a gene.

The genetic view may be satisfactory in theory, but it is difficult to imagine a real system based on nucleic acids that would have been simple enough to have formed spontaneously in a primitive environment (chapter 4; see also footnote on p. 138). Haldane (1965) attempted to draw up a simplest possible blueprint for a primitive organism (it consisted of an RNA gene specifying a single enzyme). He came to the conclusion that a primitive organism would be too improbable a structure to form according to known physical laws.

He suggested that other hitherto undetected laws would be required to account for the appearance of such a system. Perhaps Haldane was being too pessimistic. This is a technical difficulty, and it might well be overcome; but I am inclined to think that in any case it is a mistake to jump too quickly to the conclusion that the now universal nucleic acid was the original material within which evolutionary progress was written.

GENETIC METAMORPHOSIS

It seems to me likely that the evolution of *any* life form, if it proceeds for long enough from an origin, would be expected to involve at least one major switch in genetic materials. This idea, which I have called *genetic metamorphosis* (Cairns-Smith, 1966, 1968) will be the subject of the final chapter. For the moment, I will simply assume that the idea is a reasonable one—in particular that the primitive genetic material need have borne no chemical resemblance at all to the 'advanced' material DNA.

A POSSIBLE PRIMITIVE MOLECULAR BIOLOGY

If we can place no reliance on biochemical similarity between the most primitive and present day organisms, how can we have any idea about the constitution of the very first forms?

We can adopt what might be called 'the method of overlap'. There may be no single fact that will force acceptance of a particular primitive biochemistry, but there are several general considerations each of which should define more or less clearly a different area of possibilities. The system that we require should lie in the zone overlapped by all of these areas.

A good plan is to start by considering areas that are as restricted as possible; to tackle, so to speak, the most 'difficult' part of the problem first. For initial organisms the nature of their genetic material seems to be such a critical area. If our discussions so far are valid, we really know quite a lot about primitive genetic materials. We know that, like any genetic material, a primitive one has to be able to hold stably a specific pattern which is one of a potentially Archimedian number of possible patterns. Furthermore, such a pattern must be accurately replicable and it must exert an influence on its surroundings specifically tending to encourage its own propagation. The material patterns must occasionally suffer limited arbitrary changes (mutations) which are heritable. Again functionally efficient patterns must be accessible to evolution through natural selection by belonging to a reasonably continuous 'abstract landscape' with reasonably large 'catchment areas' (see chapter 6). For primitive genes relevant to the origin of life, it should have been conceivable that they could have been formed continuously on the primitive Earth through a suitable supply of material units and energy within a biologically unevolved environment. Finally, some such primitive genes should be easy to make.

We may enumerate, then, the following requirements:

(i) Information capacity
(ii) Replicability
(iii) Control of environment
(iv) *Occasional* mutation
(v) Evolutionary accessibility of functional information
(vi) Source of units
(vii) Source of energy
(viii) Ease of synthesis.

Accuracy of Crystallisation Processes

Let us start by taking points (ii) and (iv). The first organisms could have been naive and inefficient in almost every way *except* in their genetic printing mechanism because, whatever else may have been true about them, they had to be able to evolve to a more than trivial extent, and this would demand a considerable replicative accuracy. If there were several mistakes made every time genetic patterns were copied, then any 'good' pattern that happened to turn up would rather soon be rubbed out again. Indeed evolution depends on occasional mutations—but the emphasis is on the *occasional.* As Schrodinger (1944) stressed when considering the nature of 'modern' genes, it is not mutations that need explaining but their rarity. Now consider also point (viii). Where do we find accurate replication processes in the 'unevolved' world of the chemist? It is in simple crystals rather than in *simple* organic polymers that such processes are common. Indeed crystallisation processes can be fantastically accurate—that a grain of sugar has a clearly regular outline is an immediate indication of this. And the size of a grain of sugar—it is *enormous* from a molecular point of view—indicates that when sugar crystallises there is a very strong preference indeed for independent sugar molecules to fit onto a pre-existing crystal rather than start up a new crystal. The number of 'mistakes' during crystallisation is commonly far fewer than the number of 'mistakes' during ordinary organic chemical reactions. If biological reactions are an exception to this—if, for example, DNA is very accurate in its functioning—this is probably because the functioning of DNA depends on processes that are very like crystallisation processes: the selection of each new unit during copying of DNA messages must depend critically on packing considerations. Now, so far as we know, the accurate functioning of DNA depends on the pre-existence of an enzyme. Perhaps it will be found that nucleic acids, or some other kinds of organic polymers, can replicate accurately enough without such assistance. But in any case it would seem that the zone of overlap defined by (ii), (iv) and (viii) contains mainly *crystals*. I suggest, then, as a preliminary hypothesis that the primitive genes were not just *like* crystals, but actually *were* crystals of some sort.*

Direct-acting Genes

Now let us consider (iii). In modern organisms the genetic chain of command is remarkably long. To have an effect, a pattern in a DNA molecule must first be translated into RNA, then into a sequence of amino acids which may then have to fold up to give a three dimensional object—such as an enzyme. Finally the protein (e.g. enzyme) may do something useful. It is as if DNA were a government administrator in an advanced civilisation. He controls his world through his typist. He spends his time producing RNA messages—and leaves the sordid business of actually handling molecules to the proteins and the products of the activities of proteins. But primitive genetic materials could not have been so fastidious. In the early days there would have been no place for administrator-genes because there would have been no pre-existing literate civilisation. If a replicable pattern in a primitive material was to have a selective advantage, if indeed it was to have any significance at all, then it would have had to demonstrate this directly. From (iii) and (viii), primitive genes must have been *direct-acting genes*: they must have been capable of both a replicative function, like DNA, and a control function, like protein.

---

* Schrodinger (1944) suggested that 'modern' chromosomes should be regarded as 'aperiodic crystals'. This is not a bad description of DNA. I think that it will turn out to be an even better description of primitive genes.

Crystals as Direct-acting Genes?

We discussed in chapter 3 that a crucial feature of protein—probably its principle *raison d'être*—is its ability to form 'inverse images' (roughly 'sockets') of other atomic arrangements (in particular of molecules and transition states). It is largely through the extraordinary selectivity of such 'sockets' that proteins operate on their surroundings. We remarked that crystal surfaces show a similar, if less dramatic, selectivity and for a similar reason—an incomplete crystal surface contains 'sockets'. If 'plug-socket' interactions are the central events of biochemical control in modern organisms, then it is perhaps on the surfaces of crystals that we are most likely to find the primitive counterparts. To be a genuine counterpart it is necessary that the control exerted by such a crystal surface should have a genetic origin, as the control exerted by proteins has a genetic origin, and the most direct way in which crystal surface control could be genetic would be if the crystal in question were a gene—a direct-acting gene.*

Thus both the replicative and control aspects of a primitive gene lead to the idea of a crystal—to the idea that *at the core of primitive organisms there was some kind of solid state mechanism, that stood in for the complete DNA → RNA → protein system of modern organisms.*

Points (vi) and (vii) (p. 116) can be accommodated if we suppose that parts of the primitive environment were at least sometimes supersaturated with respect to the genetic crystals.

---

* A number of thinkers have suggested that crystalline mineral surfaces could have exerted an important organising effect on organic molecules on the pre-biological earth. In particular, Bernal (1951) pointed out that organic molecules would congregate in an orderly way on the surfaces of clays and that this could be expected to give rise to orderly chemical reactions. Bernal also pointed out that since quartz crystals are optically active, adsorption on quartz surfaces could possibly have provided a measure of optical resolution in local regions. Such processes could well have enriched the primitive environment—they could have provided 'food' for primitive organisms. But the processes themselves could not have been part of the metabolism of evolvable primitive organisms unless these controlling mineral surfaces were formed within the organisms as an essential part of the organisms. To be *part* of an evolvable system the specific control exerted by a surface must be genetic: it must have been inheritable from parents and be transmittable to offspring: it must result from the specific effect of some pattern which is replicable, capable of occasional mutations, and so on. An organism may pick up its atoms, its energy, even its 'orderliness' from its environment—indeed it must do so—but it can not acquire the specificity of its catalysts in this way: this bit is 'cunning' (cf. p. 64) and must have been inherited from its parent(s).

Point (i) implies an imperfect crystal—a perfect crystal can no more contain information than a book with all the letters 'a's'. We might imagine a crystal containing a particular pattern of dislocations that replicates as the crystal grows (see chapter 2). So long as the patterns are buried inside the crystal they could have little effect on the outside world, but crystals commonly break up as they grow, and in doing so may expose specific dislocation patterns that are being replicated. Now the catalytic properties of solids can depend critically on defects emerging at their surfaces. One can envisage, then, the reproduction through crystal growth and crystal cleavage of a more or less specific catalytic ability.

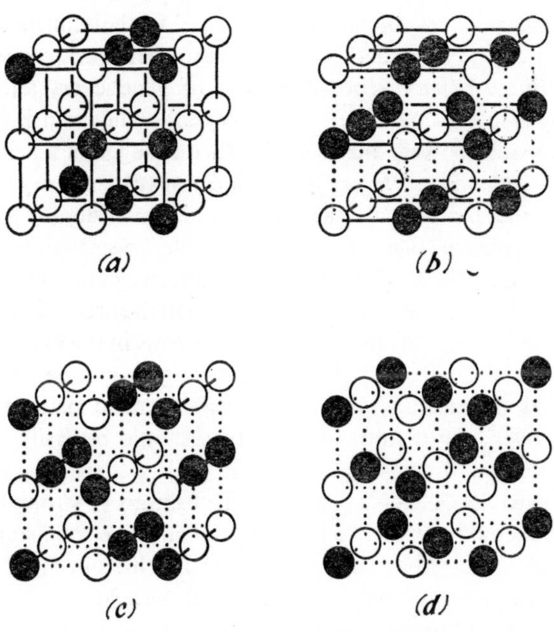

Fig. 34. Showing a small part of (*a*) a disordered substitutional crystal; (*b*) a semi-ordered substitutional crystal (disordered in two dimensions); (*c*) a semi-ordered substitutional crystal (disordered in one dimension); (*d*) is an ordered crystal. If you move along dotted lines in any of these diagrams then you can predict the colour of the next unit—it will be opposite to the previous one. Along the heavy lines there are no rules. (*b*) or (*c*) could possibly be genetic crystals.

I am inclined to think that the most effective genetic crystals would hold their specific patterns in the form of chemical, rather than physical, imperfections. In chapter 2 (p. 18) we discussed crystals (such as brass) in which alternative units can take up equivalent positions in the lattice. In such a substitutional crystal the arrangement of the alternative atoms may be completely disordered as in Fig. 34(*a*), or completely ordered as in Fig. 34(*d*). But intermediate situations are also possible in principle. We can say that (*a*) is disordered in three dimensions while (*d*) is disordered in no dimensions. Then the intermediate situations are represented by (*b*), which is disordered in two dimensions, and (*c*) which is disordered in one. Now, some 'disorder', in the sense of no simple internally visible regimentation, is necessary in a structure if it is to hold information. (Compare the 'disorder' in the sequence of bases in DNA, or the 'disorder' in a meaningful letter sequence.) So we can eliminate (*d*) as a possible kind of genetic crystal. We can probably also eliminate (*a*); such a crystal could hold information but this would be very inaccessible. Also it suggests no simple self-copying mechanism. We are left with (*b*) or (*c*); such crystals could hold specific patterns in the disordered dimension(s) and could possibly replicate these patterns through crystal growth in the ordered dimensions. The replicated patterns could then become accessible through specific crystal cleavage in the ordered dimensions. This last point requires that the kinds of bonds in the disordered and ordered dimensions should be different: they should be strong in the disordered dimension(s) so that specific genetic patterns can be retained over long periods of time: they should be easily breakable in the ordered dimension(s) so that such patterns may become accessible to the outside world. The dispositions of strong and weak bonds that could give rise to two kinds of effective solid state genetic devices are indicated by the solid and broken lines in Fig. 34 (*b*) and (*c*). In (*b*) the genetic information would be held as a 'picture', in (*c*) more conventionally as a sequence.

Do such crystals exist, and if so could they have been formed continuously from supersaturated solutions on the primitive Earth?

The Primitive Gene as a Silicate?

I have suggested (Cairns-Smith, 1966) that some rather insoluble inorganic crystal, in particular some layer-lattice silicate, might fit all the required specifications for a primitive genetic material.

According to Pauling (1960) bonds between silicon and oxygen atoms are about halfway between ideal covalent and ionic types. It is a sufficiently 'faithful' bond to be regarded as covalent: silicates are not only crystals but molecules (see later). It is often convenient, however, to regard silicate minerals as ionic crystals in which the anions are mainly large O^{2-} ions with (usually) much smaller cations fitting into spaces between them. The most important of these small cations are silicon (Si^{4+}) and aluminium (Al^{3+}). Commonly silicates are substitutional crystals with a more or less disordered arrangement of cations. This feature is common in layer-lattice silicates—particularly in micas and clays.

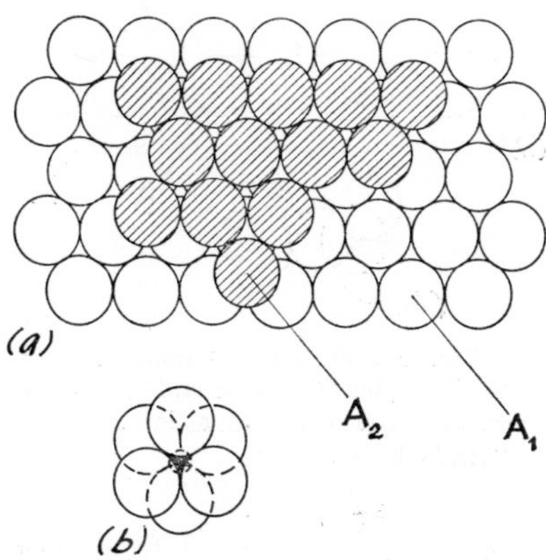

Fig. 35. Simple close packing of two layers of spheres (*a*) gives rise to 'octahedral sites' which are surrounded by six spheres (*b*).

A rough model illustrating the structure of the basic unit of mica-like silicates can be made by glueing together ping-pong balls. You first glue together a closely packed sheet of ping-pong balls (sheet A_1, Fig. 35). Then you add another similar sheet (sheet A_2, Fig. 35). Now you glue on two more sheets (B_1 and B_2) one on either side of your double sheet, only this time use somewhat smaller balls and arrange them in the form of a hexagonal net as in Fig. 36.* This four-sheet structure represents the oxygen anions in a unit mica-like silicate

* A more accurate model has been described by Radoslovich and Jones (1961).

layer. (These anions are mostly O^{2-} ions with one-third of A-sheet oxygens as OH^-). In a more complete model, small cations would be shown fitting into two kinds of niches, or sites, between the oxygen sheets. Between the two A-sheets there are *octahedral sites* surrounded by six oxygens (see Fig. 35). Between an A-sheet and a B-sheet there

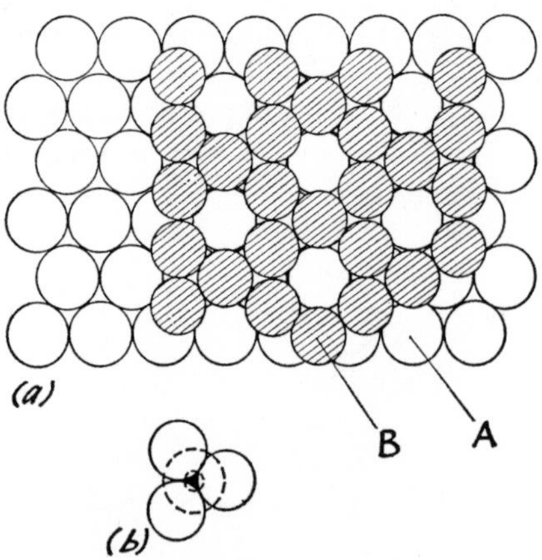

Fig. 36. A flat hexagonal net of smaller spheres may be laid on a close packed layer of spheres as in (*a*) to give 'tetrahedral sites' (*b*).

are only *tetrahedral* sites surrounded by four oxygens (see Fig. 36). A complete model of a unit mica-like silicate layer, then, would show seven sheets of ions: four sheets of large oxygens with two 'tetrahedral' (t) and one 'octahedral' (o) sheet of small cations sandwiched between them in the sequence:

$$\begin{array}{c} B_1 \\ t_1 \\ A_1 \\ o \\ A_2 \\ t_2 \\ B_2 \end{array}$$

Silicon ions may occupy all the tetrahedral sites although it is very common for a proportion of these ions to be substituted by aluminium. Octahedral sites may be occupied by a variety of ions—aluminium and magnesium are common here.

If you put Si^{4+} ions in each of the tetrahedral sites and Mg^{2+} in each of the octahedral sites, then the resulting structure will be electrically neutral—the negative charges on the oxygens will be balanced exactly by the cation charges. This is the situation in the layer silicate talc. But if some of the silicon ions are substituted by the lower charged aluminium (Al^{3+}) ions, then this will lead to a net negative charge on the unit silicate layer as a whole. Such a net charge may also arise, or be added to, through substitutions or incomplete occupancy of octahedral sites.

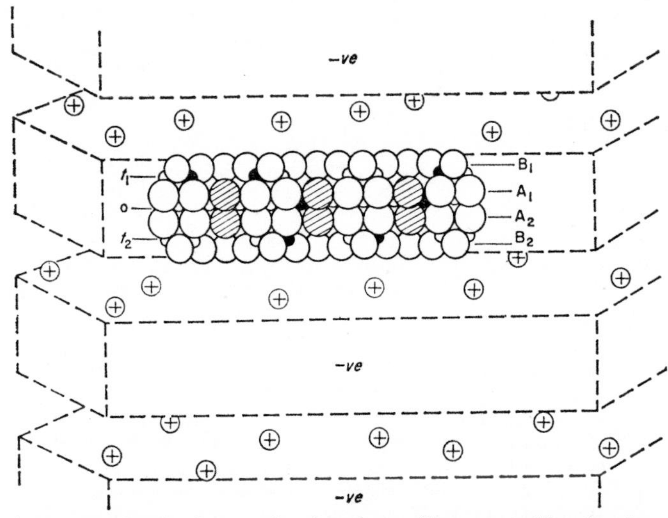

Fig. 37. Sketch of part of a three-layer silicate crystallite showing internal cation substitutions (● for ○) that give rise to negative charges in the sheets, and which may have been the elements of the primitive genetic information store.
○, o^{2-} ions; ○ and , ● intra-layer cations; , ⊕ inter-layer cations; ●, OH ions.

A complete mica-like crystal consists of a stack of negatively charged unit silicate layers interleaved with more or less mobile cations, such as potassium or sodium. The overall arrangement is illustrated in Fig. 37. The crystals can be cleaved relatively easily in the plane of the layers.

Although more complicated than our idealised genetic crystals (Fig. 34) it looks as if some such structure might fit the requirements for a primitive genetic material. Information could be held in the form of internal cation substitutions (probably most stably in the tetrahedral positions). It could be exposed through inter-layer cleavage (this can often be brought about simply by a change in the concentrations or types of ions in the environment). The next question is whether there is any indication of printing between adjacent layers. Here we can only consider a mica (muscovite) because this is the only material for which direct evidence of the kind that we need is available.

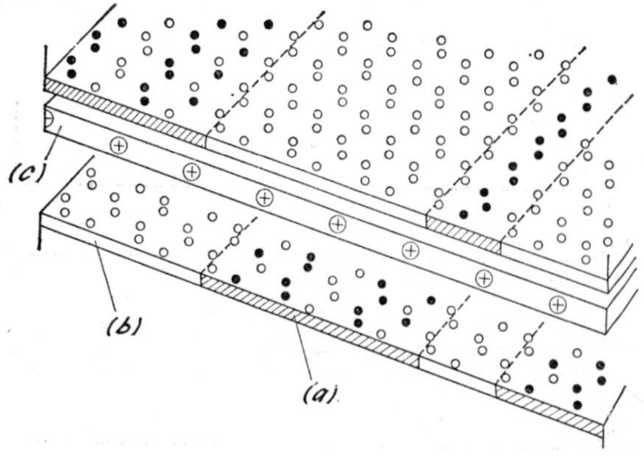

Fig. 38. Tetrahedral cations in two adjacent sheets of a single domain in muscovite, and an arrangement of these cations into parallel half-substituted (*a*) and unsubstituted (*b*) bands. Between the sheets there is an ordered array of K^+ ions (*c*) neutralising the Al^{3+} substitutions (●) either above or below. The band structure (and domain structure) in one sheet is thus copied in reverse in the adjacent sheet (Gatineau, 1964). If copying in fact occurs during the growth of this or similar silicate crystal, it could have provided an information replicating mechanism for primitive organisms. ○, Si; ●, Al; ⊕, K^+.

In micas there is one in four substitution of silicon by aluminium in the tetrahedral sites. Gatineau (1964) has studied these substitutions in muscovite in detail (from diffuse X-ray patterns). He finds that the aluminium substitutions are arranged in zig-zag parallel *rows* that are grouped into parallel *bands* in such a way that half the silicon atoms within a band are substituted by aluminium. Figure 38 shows the lower and upper tetrahedral sheets of two adjacent silicate layers illustrating an arbitrary arrangement of parallel half substituted and unsubstituted bands. This figure also illustrates a very interesting feature of the muscovite structure as worked out by Gatineau: *between two adjacent silicate layers an unsubstituted tetrahedral band comes immediately above a half substituted one and vice versa.* The proportion of one in four for tetrahedral substitutions in mica is thus neatly accounted for as arising from a vertical pairing of unsubstituted with half substituted bands. The arrangement suggests a possible printing mechanism: it suggests that a pre-existing arbitrary banding pattern may be copied in reverse through the formation of a new layer on a pre-existing layer during crystal growth.

A mica crystal consists of small domains within which the bands are parallel. The direction of the bands changes abruptly at domain boundaries. There are three principle orientations at 60° to each other—the whole effect is rather like a crazy-paving of striped stones with the stripes on most of the stones orientated at 'six o'clock', 'twenty to two' or 'ten to four'. As the whole crazy-paving pattern is repeated again and again between the layers, we can imagine 'information' in the form of a specific banded crazy-paving pattern being printed off repeatedly as the crystal grows.

We cannot say that the semi-ordered structure of muscovite *proves* that substitution patterns are printed during crystal growth: all that we can say is that mica shows general structural features that we would expect of a genetic crystal.

The same factors that create the semi-ordered arrangement of substititions in mica* may very well operate in other silicate minerals of similar constitution. Indeed, we should not at this stage limit our thinking to a particular group of silicates: I would guess that many silicates will turn out to have semi-ordered features of possible proto-biological significance. Kaolin, antigorite, montmorillonite and tobermorite represent some other classes of layer silicates: we should add also fibrous silicates, e.g. amphiboles, as possible examples of our second type of genetic crystal (see Fig. 44). I will not, however, pursue such particular speculations further: I will try instead to

---

* Gatineau suggests that the inter-layer ordering results from requirements of local balance of electric charges: a half-substituted region will have an excess of negative charge that will just balance completely the charges on all the adjacent interlayer potassium ions. For the complete sheet of inter-layer potassium ions to be locally neutralised, then, there must be everywhere a half-substituted band *either* above it *or* below it. The situation of lowest energy results, then, if the substitution arrangements in the upper and lower tetrahedral sheets are strictly complementary. (For a further discussion of such ordering processes see Cairns-Smith, 1966.)

indicate just how important the confirmation of a 'printable' silicate would be. Because it seems to me that it is this question of printing that is crucial: if this could be demonstrated in any one of a number of silicates then we would have the ground material for a plausible proto-evolution available for direct study in the laboratory. (In other words, I think that all the other requirements listed on p. 116 would be satisfied relatively easily. I will now try to justify this.)

Assume Accurate Printing—Then What?

We are familiar with mica as an obviously crystalline and rather striking material, but it is almost certainly exceedingly minute crystals—crystals of colloidal dimensions—that we should be thinking about. Quite probably some material that we would describe as 'clay'. Viewed from the other end of a garden spade, clay seems to be a rather formless kind of stuff. But if you look at it more closely—under the electron microscope—its underlying layer structure becomes evident: it may appear like a mass of wet crumpled sheets (montmorillonite, Plate IV), or likes piles of roughly hexagonal slates (kaolinite, Plate V), or again like folded ribbons (rectorite, Plate VI), or rolled up tubes (halloysite, Plate VII), or in various other forms corresponding to different specific types.

Are Silicates Strong Enough?

In particular it is worth noting that, at the microscopic level, a clay is typically a flexible, tough sort of material. Some of the sheets shown in the electron microscope pictures (Plates IV–VIII) are only one or two unit layers thick—yet they can be folded without breaking. The picture of rectorite is particularly striking: the thinnest of the ribbons is only two unit layers (8 oxygen atoms) thick. This flexibility and toughness emphasises that unit silicate layers should be regarded as two dimensional *molecules* as well as crystals.*

Are Silicates too Strong?

It is often stated (e.g. Bernal, 1965) that silicates could have no serious role within organisms operating at ordinary temperatures because the Si—O bond is too strong: it could not partake in the

* Rex (1966) has extracted mica crystallites from a sandstone within which they formed. Many of the crystals were only a few unit layers thick, as shown in electron micrographs, yet they survived the extraction procedure.

continual covalent changes supposed to be characteristic of life. But there is at least one place for a highly stable covalent arrangement of atoms within an organism—right at the centre—as a gene. It is not necessary that a gene should undergo covalent changes within itself after it has been formed. Nor is it necessary that it should make and break covalent bonds with its surroundings in controlling its surroundings. The genes in modern organisms may cause covalent changes in and between molecules outside themselves, but they do this through interactions involving secondary forces. Right down the 'chain of command'; in the formation of an RNA reprint from the DNA master: in the ribosomal manufacture of a specific poly-amino acid: in the catalytic or other effect of the resulting protein; the *control* is effected almost exclusively through non-covalent interactions. Outside organisms, also the best 'control devices' are often chemically inert materials—think of catalysts like platinum, nickel, alumina or, for that matter, clay.

Even during the replicative synthesis of a gene, it is not necessary either for the 'old' or the 'new' gene to be chemically reactive. What the 'old' gene must do is to orient new units, and it can best do this through non-covalent 'packing forces'. It is only the new *units* that must be chemically reactive to form covalent bonds *with each other*. And the stronger these bonds are the better—they are never going to have to come apart again.

If we can assume that on the surface of the primitive Earth there were rocks that had been formed internally at high temperatures and pressures—as is true today—then some of these rocks could have provided not only material units but also energy for the proposed genetic crystallisation processes. We do not usually think of rocks as *fuels*, but many do indeed have energy locked up in them. Rocks that have been formed at very high temperatures and pressures will in general only be stable at such temperatures and pressures: if they appear at the surface of the Earth they will usually be metastable and tend to change into something else. For examples, feldspars (constituents of granites) in contact with solutions of inorganic ions are often metastable with respect to layer silicates at ordinary temperatures. Feldspars may thus tend to change into mica, montmorillonite, or kaolin (together with quartz) depending on the exact conditions prevailing. Such changes occur during the weathering of rocks—water providing a means of mobilising the units (e.g. SiO_4) and allowing them to reform in more stable arrangements. Pedro

(1962) has demonstrated that clay minerals can be formed by continuously washing rock samples (e.g. granite) with hot water.

Rock weathering is a complex subject about which there is much dispute, but two general points seem clear: (i) decomposition of silicates can lead to dilute solutions of silica and other ions (Eitel, 1966), and (ii) such solutions can give rise to clays at quite low temperatures. This second point has been demonstrated by Henin and his collaborators, who have synthesised in the laboratory a considerable number of different clay types from very dilute solutions (Henin and Robichet, 1954; Caillère, Oberlin, and Henin, 1954; see Plates VIII and IX). These syntheses were carried out at 100°C and at room temperature. Typically, decigram amounts of product could be made conveniently in a few weeks. Wey and Siffert (1962) made a magnesium montmorillonite from very dilute solutions of silica and magnesium ions slightly alkaline conditions. This product formed overnight at room temperature. A similar product is shown in Plate X. Hectorites are also relatively easily formed by heating slurries of silica and magnesium hydroxide, particularly in the presence of lithium fluoride (Granquist and Pollack, 1960).

It seems reasonable to suppose, then, that the earth produces now, and could have produced in the distant past, situations in which layer lattice silicates could have crystallised continuously from solution at a reasonable rate.

What can Layer Silicates Do?

In considering the question of possible modes of environmental control ((iii) p. 116) we may note again that various organic molecules were almost certainly present on the primitive Earth and that clays have a dramatic ability to adsorb organic molecules (as well as inorganic ions) on their surfaces and edges. Highly organised 'organo-clay complexes' form with alcohols, amines, amino acids, proteins, urea, pyridine and many other kinds of molecules. The arrangement of a mixture of an amine and an alcohol as deduced by Weiss (1963) is illustrated in Fig. 39. (The amine has taken the place of the inorganic inter-layer cations and the spaces left have been filled up with the alcohol.) Polypeptides of different but discrete sizes can be formed by reaction of suitably activated amino acids on montmorillonite (Paecht-Horowitz, Berger and Katchalsky, 1970). Normal chemical properties may change when organic molecules are fixed by clays. For example, benzene forms a complex with copper (II) within montmorillonite (Mortland and Pinnavaia, 1971).

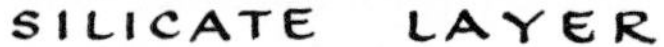

Fig. 39. Packing of the alcohol $CH_3(CH_2)_{11}OH$ and the amine $CH_3(CH_2)_{11}\overset{\oplus}{N}H_3$ between layers of mica-like silicates. It is known that the ratio of alcohol to amine is fixed by the clay (Weiss, 1963) and one might speculate that the arrangement of the alcohol and amine molecules is also controlled by the arrangement of charges in the silicate layers. This speculation is indicated in the diagram.

A visible piece of clay consists of a vast number of partly stacked, partly crumpled, and partly folded layers. The overall physical *consistency* of the clay will depend on the way in which these particles interact with each other. The surfaces of the layers carry negative charges arising from internal substitutions: the edges carry positive charges arising from exposed internal cations, thus in addition to layers tending to stack on top of each other, via interlayer cations, they may tend to associate at right angles (i.e. not only face to face but edge to face). Figure 40 illustrates some possible modes of interaction between clay particles. The extent and balance of such interactions can determine, for example, whether the clay is runny or gelatinous or heavy or mushy. Also, quite small amounts of organic and other components can dramatically affect the rheological

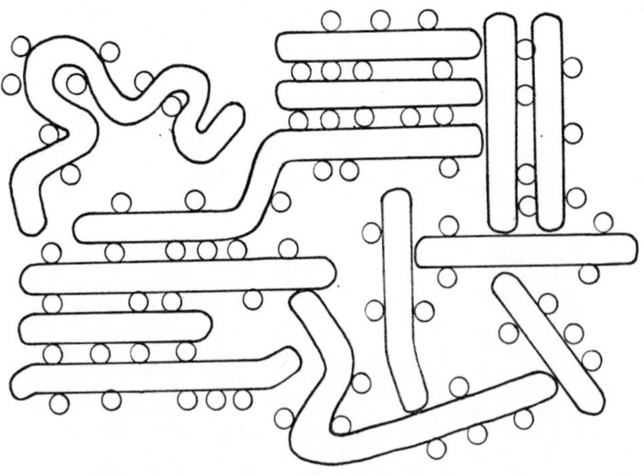

Fig. 40. Edge-on view of various kinds of interaction between silicate layers.

properties of a clay either by helping to tie the clay particles together ('crosslinking effect') or by keeping them apart ('protecting effect'). Indeed, the same additive may have these opposite effects at slightly different concentrations (van Olphen, 1963). Without at the moment being more specific, it is not unreasonable to suppose that differences in internal substitution patterns within clays could have an effect on rheological properties (gooeyness) through effects on organic molecule adsorption and the other kinds of interactions which we have discussed. Indeed, there are many other possible effects of more specific patterns of substitutions (footnote on p. 139), but this is enough to be going on with: we can already begin to imagine a possible 'origin of life'.

An Easy Way of Life?

We suppose that on the primitive Earth there was land including rocks that weathered to create supersaturated solutions from which clays crystallised from time to time. More specifically, we might imagine such solutions percolating through porous rock or a bed of sand with clay crystallisation being nucleated more or less at random—say on the surfaces of the surrounding rocks—giving rise to arbitrary substitution patterns. In at least some of these clays, however, these initial arbitrary patterns are subsequently printed off again and again through crystal growth. We see this as the pre-condition for life: genetic crystals in a supersaturated solution automatically printing off any pattern that they just happen to hold. There is no need for elaborate pre-formed machinery—no need for a printing press—to effect the pattern replication: it is, so to speak, provided by the environment as a free service: it cannot help but happen as the crystals grow.

Now if patterns get printed equally well whatever they are, there is no scope for evolution. There is a limited scope if some kinds of pattern, for some reasons of crystal geometry, are printed more quickly than others—perhaps very symmetrical ones might be favoured. But this is still hardly enough. This way of life is too easy to be much of a life at all. We must introduce a problem—just a little one. We must consider a situation in which some patterns will cope better than others: in which they will exert some kind of control on their environment tending specifically to improve their chances of surviving and replicating. There is no need for this control to be through the catalysis of elaborate networks of chemical reactions—indeed it must be simple if it is to arise by chance and it need not be catalytic at all.

Maybe a suitable problem will hit our system with the next shower of rain. The water flow becomes more vigorous, perhaps, and ion concentrations change. The problem is simple and vital—to stay in the stream, not to be washed away. Now here is a chance for the different patterns which are being blithely printed off to show their mettle. Let us think how they might cope.

THE STORY OF SLOPPY, STICKY, LUMPY, AND TOUGH

Somewhere in the stream, let us say, there are four particular distinct patterns which have printed themselves millions of times to create four small regions in which the physical consistency of the clays are different (Fig. 41)—such differences arising from differences in the patterns as discussed earlier. On the left we have a clay with a rather loose, open consistency—call him 'Sloppy'. Next, in another region,

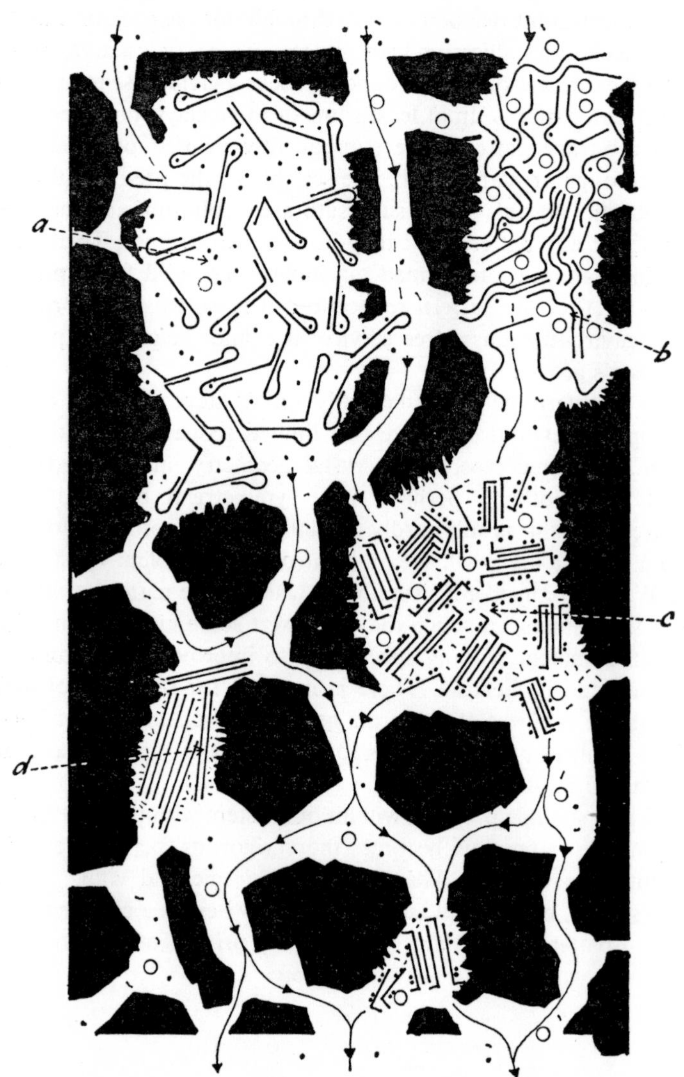

Fig. 41. A hypothetical 'origin of life'. We imagine clay platelets forming replicatively from solutions flowing through porous rock or sand. Different substitution patterns give clays that have different folding habits and which stick to each other in different ways: they also adsorb and entrain different proportions of molecules in the environment. Thus some patterns survive and replicate more effectively than others (see text).

we have 'Sticky'—maybe he has a pattern that tends to select rather a high proportion of sugar-like molecules from the solutions flowing past; in any case, his stickiness lets him hang on quite well to the rock. Further down (*c*) there is 'Lumpy', whose clay platelets print off patterns that give rise to local coagulations like a badly made custard. The final member of our humble quartet is 'Tough' (*d*). He hangs grimly onto the rock—one suspects that his charge pattern happens to correspond roughly to the characteristic surface of the local rock.

Before the rain came Sloppy was winning easily. With his diffuse structure there was plenty of scope for the necessary silicate units to diffuse into place: clay synthesis was relatively rapid and the secret of Sloppy's sloppiness was being promulgated very effectively. Tough, at the other extreme, was slowly grinding to a halt as his own impermeability blocked his further progress by diverting the local flow of 'foods' away from himself. Both Sticky and Lumpy were growing at a moderate pace and there was not much to choose between them.

Then came the rain. Sloppy was quickly swept away. He was never seen again. Sticky hung on quite nicely, and so did Tough; but the increased flow rate did not do either of them any good since the solution was no longer supersaturated. As it turned out it was Lumpy who was best adapted to the situation. He broke into pieces which re-established themselves farther downstream—to carry on the good work within a wider territory when conditions again favoured clay synthesis: to make more and more of the lumpy material by printing off countless copies of the pattern that caused the platelets holding it to associate with each other, and with environmental molecules, in a suitably lumpy way.

CATCHMENT AREAS

Of course, the story of Lumpy is just a story. Even if some such evolution did happen on the primitive Earth it probably had an unhappy ending, it would probably have been one of countless false starts. There are many possible niches even for the kinds of organisms that we have been thinking about, and many possible ways of life within each niche—many possible tricks that might enhance the prevalence of a certain type of crystal defect pattern. The point of the story was to illustrate just how simple such tricks could be. In terms of the model that we introduced in the last chapter, some broad property like stickiness or lumpiness could have represented a wide 'catchment area' in the relevant evolutionary landscape: to determine such a property is not likely to require the specification of a particular substitution pattern or even one of a small set of possible patterns. More likely the specification would be quite vague, such as having a rather high density of charge at one end of a longish platelet and a somewhat smaller charge density at the other end—some specification that could easily have been conformed to by chance given a few million trials. *Any* of the vast number of patterns conforming to this general specification could have caught on.

PROSPECTS FOR FURTHER EVOLUTION

But the evolutionary landscape must not only be accessible it must be interesting. There must exist 'local minima' towards which systems could roll. A system, that is, should be capable of further, more ingenious, adaptations to its environment through natural selection. There is no immediate difficulty in principle here. One has only to begin thinking about a naive system (like Lumpy) to see possible improvements that could have arisen stage by stage through natural selection operating on occasional errors in printing. If a mutation occurs in some region of the material that happens slightly to improve the effectiveness of the break-up process, then lumps derived from this region will eventually have more progeny. To begin with, Lumpy would perhaps have picked up only rather general kinds of molecule from the environmental solution—later he might become more fussy. If he happened to increase his affinity for phenols and carboxylic acids—particularly if he could select specifically diphenols and polycarboxylic acids—then he might find himself on a very good thing indeed, because these kinds of molecules can increase the solubility of silica—they might thus act on the surrounding rocks to provide an immediate source of food.

Perhaps, like the amoeba, Lumpy would eventually run out of ideas: he would reach a situation in which no improvement limited enough to be accessible to chance would be possible. He would, like the amoeba, be in a deep local minimum, well adapted to his environment (and one might almost say thoroughly conservative in his outlook). But this would not necessarily be the end of the story. We must remember the second major factor in evolution—the tendency for hereditary systems to move, not only downwards on the evolutionary landscape towards local states of optimum biological efficiency, but also outwards towards wider fields of possibility (chapter 6, p. 107).

One might think of a region in which genetic crystals can form rather easily with gradually less and less clement regions adjacent to it. Having originated and having started to evolve through natural selection in the 'easy' regions, organisms would then tend to colonise the more 'difficult' regions. This would not be through obtuseness, or any conscious spirit of adventure, but in the first place simply by chance. By chance some individuals would find themselves in the less attractive surroundings. Maybe they would not thrive very well there, but provided they could survive and reproduce at all the resulting population would be subject to natural selection which would tend to adapt them to the 'difficult' surroundings. New subtleties and modifications, that would have been pointless in the original 'easy' region, might then be evolved making subsequent, even deeper inroads into even more 'difficult' regions possible—and so on. The organisms eventually found colonising the most 'difficult' of the regions would on the whole be the most complex and interesting—we would probably say the most 'evolved'—as today's animals on land are, on the whole, more complex and interesting than their more conservative cousins who stayed in the sea.

An organism must have an efficient hereditary machine, however, to be pushed very far by these evolutionary factors. It is unlikely on quite general grounds that a hereditary system simple enough to arise by chance in a non-biological environment would also be particularly effective as a vehicle for prolonged and elaborate evolution. A direct-acting gene may be a good idea as a starter, but an indirect system is evidently more effective in the long run. It is not difficult to see why. With an indirect system, such as the modern one, the two quite different basic functions of replication and general environmental control are taken by two quite different molecules, DNA and protein, each a specialist. You *might* find a single material that happened to combine these two abilities optimally—but it would be very unlikely. A direct-acting gene is almost certain to be a compromise: just because it can perform two distinct functions it is less likely to be able to do either of them very well. It is, so to speak, an amateur all-rounder: DNA and protein are professionals. By the same token it is unlikely that a material able to form the genetic basis of an initial organism would be the best suited as *either* of the components of a composite system. The only grounds for supposing that the primitive direct-acting gene was either like DNA or like protein would be if there were no easy evolutionary mechanisms available for changing over from one genetic material to another. Then indeed organisms would be for ever stuck with something like the system with which they were able to start. In the next chapter I will try to indicate that a radical switching mechanism exists in principle and was probably used in fact.

FURTHER READING

OPARIN, A. I. 1957. *The origin of life on the earth.* Oliver and Boyd, Edinburgh. An intermediate account of the major theories on the origin of life up to the mid-fifties together with a detailed discussion of the author's own theory.

CLARK, F. and SYNGE, R. L. M. (Eds.) 1959. *The origin of life on the earth.* Pergamon, London. A massive collection of papers. Fox, S. W. (Ed.) 1965. *The origins of prebiological systems.* Academic Press, New York. A useful guide to possible modes of synthesis of 'biochemicals' under pre-biological conditions. SHNEOUR, E. A. and OTTESEN, E. A. 1966. *Extraterrestrial life.* National Academy of Sciences, National Research Council, Washington. This is an anthology of papers and includes several major ones on the origin of life (e.g. Haldane, 1954). Excellent bibliography. See also C.S.P.s Nos. 2 and 17 by FIRSOFF, V. A. 1967 and MARQUAND, J. 1968. Oliver and Boyd, Edinburgh.

8

Genetic Metamorphosis

By 'genetic metamorphosis' I mean a hypothetical major change in genetic materials taking place under the influence of the twin adaptive and dispersive factors of evolution. I do *not* imagine this process as any kind of rewriting of genetic instructions into a new material. Neither do I see it as a gradual change through a long succession of intermediate types. The mechanism that I suggest is a kind of takeover. It can perhaps be outlined most easily in the form of a commercial allegory.

John Clay founded The Company and was a rugged self-dependent sort of fellow who at first did everything himself and kept things going well enough. Then, from time to time, he brought in various assistants to help in minor matters about the office—making the eat and sharpening pencils. Some of these newcomers turned out to have other uses and, in the protection which The Founder had established, thrived themselves while contributing to the continued survival and extension of the whole show, more and more efficiently and in wider and wider territories (and using all kinds of cunning ideas that John Clay had never thought of and in any case would have been unable to put into effect by himself). But there are dangers in bringing in new blood: it may find that it can do without the old. In a world in which there was no respect for seniority; where those personal qualities that had made The Founder such a success in a rough uncivilised world were no longer particularly relevant; where the civilised attribute of effectiveness within a pre-organised community was now what counted—under these circumstances The Founder, although not definitely asked to resign, found himself left out of some of the more interesting ventures. (Perhaps it was his need to feed on stones that was rather a drag.) Eventually the civilising process which he had himself made possible came to destroy the opportunities for his own survival.

Now let us re-consider this story more formally in terms of subprocesses that might be expected to follow as consequences of the evolution of an initial system like that envisaged in the last chapter.* For convenience we will consider these sub-processes under separate heads.

LOCAL MULTIPLICATION OF GENES

Primitive genes would have multiplied in a largely uncontrolled way to create more or less extensive regions containing vast numbers of closely similar copies. An individual primitive gene would have quickly become a *crowd.* As soon as this had happened the normal environment for each of the individual members would contain other individuals like itself. Mutations would then be selected that tended to increase biological efficiency within this situation. The members of the crowd would become adapted to living with each other, e.g. mutations would be selected that produced efficient modes of interaction between the particles (compare the story of Sloppy, Sticky, Lumpy and Tough). The crowd, that is, would become a *congregation*—a co-operative assemblage of similarly functional units.

DIVERGENCE OF GENE FUNCTION

In a naive way we might imagine that when a mutation occurs within a congregation, then either the old gene or the new modification will

---

* The arguments will be more thinkable if one imagines a particular possibility, e.g. the clay system discussed, but they are intended to be more general than this. We see the initial genetic material as being in any case 'rugged' and 'direct-acting'—something quite different from DNA. Indeed DNA looks like a poor starter from several points of view. It is a rather fragile molecule on its own, being easily broken by mechanical stresses, e.g. when a solution of it is stirred (contrast the 'rugged' rectorite shown in Plate VI). Also it is very easily damaged by ultraviolet radiation—a particular hazard on the primitive Earth where there was probably no oxygen in the atmosphere and hence no ultra-violet filter of ozone. Again it would have been liable to undergo covalent chemical reactions with an environment rich in reactive organic molecules. Again the units required to make nucleic acids, even if they had been around on the primitive Earth would usually have been there in an impure form: in particular there would have been impurities of similar chemical constitution to the units, which would have been likely to have confused the replication processes. That a nucleic acid is built from only one of the optical isomers of its units is a particular difficulty here. Indeed one can imagine situations in which these difficulties might not apply—particularly if you think of them one at a time. But for all these difficulties to have been overcome at the same time in a particular situation over a long period, and overcome sufficiently well for replication processes to be really accurate—this seems highly unlikely. A DNA-like molecule *might* have been the initial genetic material, but it is hardly a first choice if there are other possibilities.

eventually 'win' and displace the other. But there might sometimes be a positive advantage to a congregation (and hence to the individuals that make it up) if its members were *not* all the same. In some cases a suitable mixture of a new and an old gene might be more efficient than either on its own. For example an ideal physical consistency might be more nearly realised with a mixture of small and large particles. Later mutations in the small particles might then lead to more particular subfunctions—e.g. collecting a particular kind of organic molecule from the environment which had a general effect of promoting clay synthesis. An ideal balance of numbers between such divergent genes would tend automatically to be maintained because regions in which a suitable balance happened to exist would grow fastest and continue to do so as long as the balance remained optimal.

CO-ADAPTATION OF GENES

With divergence of gene function the congregation would be on the way to becoming a *community* in which the overall biological function of surviving and reproducing would become dependent on subfunctions performed by a variety of specialist structures.* Gene A, let us say, can survive and replicate on its own, but it does so more efficiently (or can do so in a new environment) in conjunction with gene B which performs some special service. But B has now changed

---

* There is an interesting way in which elaborate structures could in principle be specified by a charge pattern on a flexible sheet: it is reminiscent of the Japanese art of making toys and decorations by folding paper. Two adjacent layers in mica stick well together because their charge patterns 'key' with each other (Footnote, p. 125). Without such keying only about 50% of the interlayer potassium ions would have their charge balanced locally (by one negative charge *either* above or below): the rest would be unbalanced (either by having two negative charges in their vicinity or none). The 'keyed' orientation of two such layers holding an unsymmetrical—e.g. 'crazy-paving'—pattern will represent a unique low energy situation. Now think of a single independent layer with different regions on it which key with each other if the sheet is folded in such a way as to bring these regions adjacent to each other. This could provide a way of building a specific folding arrangement into a two dimensional charge pattern—rather as the specific folding of a protein is built into its one-dimensional structure. Furthermore, at particular corners, edges, or crevices within such a specifically crumpled sheet there might well exist special grooves, lattice strains or vacancies with catalytic properties. (Clays which have been suitably maltreated to introduce lattice imperfections are used in industry in vast quantities, as catalysts.)

the immediate environment of A, and this may favour a new modification of A, say A′, which is now selected:

$$A \rightarrow A+B \rightarrow A'+B.$$

There is now no guarantee that A′ could survive on its own since it evolved latterly in an environment that did not require this. B, although starting as an 'optional extra', might thus become a necessity. Similar processes could be repeated with genes C, D, etc. leading to a more or less strongly interdependent community. (Leading, that is, to gene-butchers, gene-bakers and gene-candlestick-makers in the place of the original independent savage.)

ELABORATION OF A COMMUNAL PHENOTYPE

Direct-acting genes are arguably already 'phenotype' in so far as the pattern which they replicate is immediately functional, but in any case they are likely to operate through the acquisition of additional non-genetic components picked up from their environment (and possibly subsequently altered). The environment of any member of a community of genes will thus contain not only other genes but also associated phenotype. Similar arguments to those already given would lead to the conclusion that these genes would evolve in such a way that they tended to become dependent on the phenotype created by the community as a whole. At this stage we might say that the community was becoming a *civilisation*, i.e. strongly dependent on the products of the activities of specialists.

GENE ADDITION

Apart from alterations to existing genes, new inheritable functions could arise every so often 'from scratch'. That is, some more or less completely arbitrary pattern in a piece of genetic material might just happen to contribute to a pre-existing community. As with the original arbitrary pattern in the original primitive gene, this one would amost certainly belong to a relatively large set of functional patterns: like the primary starter gene its function would probably be something simple and vague—such as increasing general stickiness, providing fibrous material, or catalysing some fairly broad category of reactions. *Unlike* the original primary starter, however, such 'secondary starters' would have a pre-organised community to which to contribute: if they could simply dot an 'i' or cross a 't' within this community they could be selected for. It is much easier to be effective, and there are far more different ways of being effective, within an organised community than out in the open. It is characteristic of any community as it evolves and becomes more specialised in its parts that it creates further opportunities for further specialised functions. (Invent candles and, without meaning to, you have created a niche for candlestick-makers.)

ESTABLISHMENT OF 'HETEROGENETIC' COMMUNITIES

A key point to notice about added genes—'secondary starters'—is that *they need not be made of the same material as pre-existing genes.* The genetic units in the kind of community we have been discussing would stay together, not through structural similarity, but through functional interdependence. (As grass, sheepdogs, sheep and shepherds are often found in each other's vicinity not because they look alike but because they are useful to each other.) Particularly in the early days, with direct-acting genes, the kinds of functions which the genes could perform would be severely limited by the materials out of which they were made. To take a particular example, there may be many clever things that can be done with clay platelets, but the subfunctions available to the community would be more varied still if other direct-acting genetic materials such as fibrous minerals or, later, organic polymers were also included in the community.

Notice particularly that factors operating in favouring secondary genetic materials would no longer be quite the same as for primary materials. A pre-organised community would not only provide a wider range of sometimes rather easy niches for added genes, but it would also provide opportunities for novel genetic materials that would have been unable to operate originally. This can be illustrated by a particular speculation. Suppose that a silicate-based community developed a common phenotype containing a rather high concentration of just a few specific purines and pyrimidines and only one of the optical isomers of ribose—these materials having some use within the quite highly evolved silicate-based community (possibly as components of energy carriers). This would increase the probability of an RNA-like material arising to perform in the first place some suitably trivial sub-function.

GENETIC EXTENSION

The evolutionary potential of a new genetic material would depend, not on the importance of the original reason for its incorporation—to begin with it must have been an 'optional extra'—but on the variety of functions that it could eventually perform. We might guess that the modern system started with (something like) RNA, because, of the three polymers used in this system, RNA is the nearest to a direct-acting genetic material (RNA *can* replicate—in cells infected with RNA-viruses: and it *can* form protein-like structures—transfer RNA's and ribosomal RNA). Perhaps in its first role it would not have seemed very impressive. If RNA eventually inherited the Earth it did so because it was capable of a far reaching extension in its action. Somehow it learnt to make protein.

The original function of RNA, that gave it a selective value within a pre-existing heterogenetic community, might have been purely mechanical—as a rather fancy kind of fibre, perhaps, or as some kind of molecular skeleton. At first, perhaps, the material was synthesised non-replicatively, and the base sequences were random giving rise to arbitrary patterns of self-twisting. Nevertheless, provided this new stuff was of some use to the community, it would be kept, and mutations in pre-existing genes that favoured its synthesis would be favoured also.

A particular point about RNA among the various polymers manufactured by primitive gene communities would have been that here there existed an alternative mode of synthesis—replication. This could have had significance for the gene community even although it was not particularly accurate—the replication in the first place need have had no genetic role. All that was necessary was that (rough) replicative synthesis of RNA should have been a good idea from the point of view of the pre-existing community—perhaps because RNA tangles would more closely resemble each other and thus associate together in a more coherent way. If rough replicative synthesis is a good idea then accurate replicative synthesis might well be a still better one. So the community might evolve improved means of replicating its RNA—not deliberately to make genes out of it, but for a purely phenotypic function such as specific fitting together of different RNA tangles (i.e. it would not matter *what* the sequences in the RNA's were so long as they were complementary to each other). But having evolved an accurately replicating structure the community would find that, whether it liked it or not, it had also produced a new genetic material.

We might imagine now mutations occurring in this new replicable material producing modified patterns of tangling and sometimes giving rise to particular crevices that happen to bind one or a few amino acids making it easier for these to join into short peptides of a composition that is somewhat variable, but nevertheless vaguely controlled by the RNA tangle. These crevices are selected for and perfected, because the peptides themselves are useful to the community—in some quite unsophisticated way. The specific crevices are then further moulded by natural selection to make longer and more accurate peptides: not because the community 'knows' that only by perfecting a system for accurate specification of amino acid sequences will it ever be able to make a decent enzyme, but because the general ability to make longer peptides more accurately would have a general selective value even in the early stages. There are all kinds of things that proteins are useful for: many of these functions would not require great accuracy of specification but would be improved nevertheless with improved accuracy of specification (think of a sticky anchor material, or a fibre connecting different parts of the community). There would thus be a continual selection pressure for improved protein synthesising machinery. As this machinery improved, so more and more of the general functions that we now associate with proteins might be discovered. Enzymes, I think, would have been a late discovery depending on the prior evolution of quite sophisticated RNA apparatus—something like a self-replicating ribosome—able to tie together long amino acid sequences with considerable accuracy. *Then* life could have restarted, so to speak, on a new super-efficient basis. *Then* nucleic acid would have become a viable proposition as the sole genetic material of a gene community. Life, which had started with a homogenetic community based on a single, primitive, direct, rugged, 'starter', could again become homogenetic—only this time with a single system that was an advanced, indirect, delicate 'evolver' (see Figs. 42 and 43).

GENETIC MODIFICATION

Figure 44 summarises a possible mechanism for the establishment of DNA as the genetic material. This kind of process, in which a

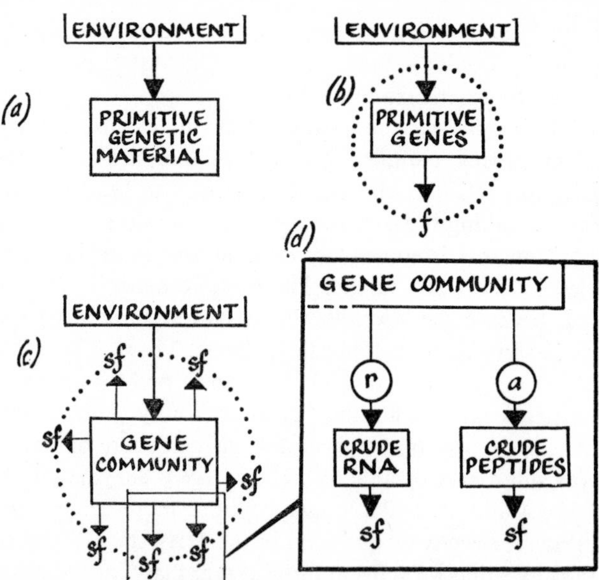

Fig. 42. (*a*) The primitive genetic material we see as having been formed continuously from the environment. (*b*) Every so often a replicable pattern appears which has some simple function (f) tending to preserve the genes holding it (this self-preservation is symbolised by the dotted circle). (*c*) Overall function subdivides into co-operatively self-preserving sub-functions (sf). (*d*) This is a close look at one such hypothetical community of genes: one of the subfunctions depends on the synthesis of a crude RNA-like polymer and another on the formation of crude peptides with virtually no sequence control. We see these as two of many similar materials used in such primitive gene communities. These are particularly plausible because the primitive environment probably produced various molecules like RNA precursors and protein precursors which would thus have been available to the primitive gene communities for incorporation in their common phenotypes (i.e. sets of molecules (r) and (a), or molecules out of which (r) and (a) could have been formed were probably available).

given genetic material is changed for another very similar one, does not seem to represent a major difficulty, particularly since the discovery that DNA can be synthesised from an RNA template (Temin and Mizutani, 1970; Baltimore, 1970). It has been suggested a number of times (e.g. Haldane, 1965) that RNA preceded DNA. One may call such a process *genetic modification*. (Such a process could have complicated an RNA takeover mechanism—the progenitor of the modern system need only have been *like* RNA).

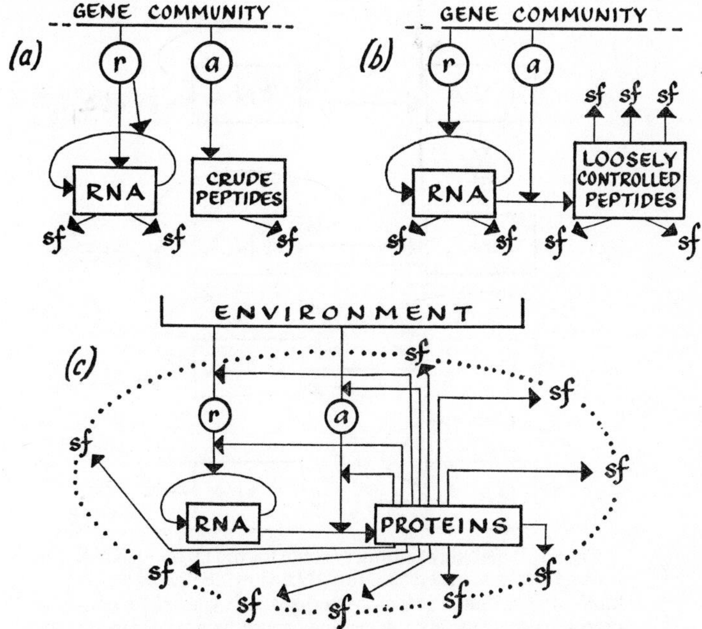

Fig. 43. We see the modern genetic machinery as having arisen within the phenotype of a primitive gene community. (*a*) RNA replication discovered as preferred alternative to non-replicative synthesis for some non-genetic function: as this replication becomes more accurate, natural selection can operate on RNA itself to elaborate new specific subfunctions. In particular (*b*) some of the peptide synthesis takes place in association with the specifically tangled RNA chains leading to a broad measure of control on the kind of peptide produced. This allows now a range of somewhat different peptide subfunction. (*c*) This is very much later. Really accurate machinery for protein synthesis having been evolved the whole system is 'brought up to date' by using protein catalysts everywhere (as well as performing most of the essential subfunctions either directly or indirectly with protein). Pre-existing gene community no longer needed to provide 'shelter', 'food', and 'jobs'. Matter and energy now obtained directly from the environment (cf. Fig. 42).

CELLS

We have been thinking of early gene communities as containing millions of genes of many different kinds, and made, perhaps, out of several different materials. We should consider for a moment processes that might have lead to the organisation of such a community into cells. The first stage in such a process might be the

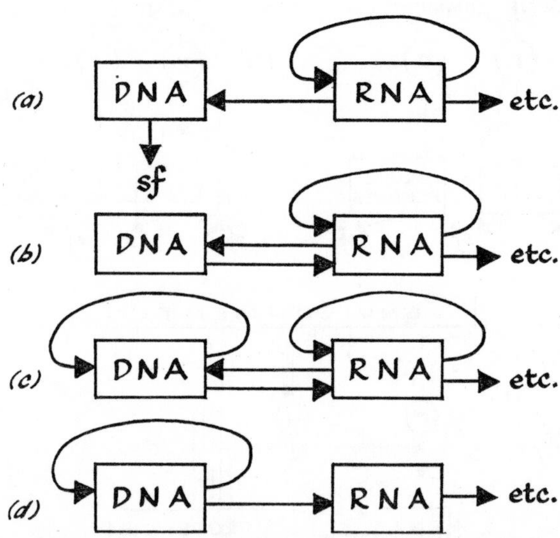

Fig. 44. Possible evolutionary route for the DNA → RNA limb of the modern system. (*a*) DNA is made first as a new kind of functional polymer—perhaps to make rigid connector pieces. (*b*) DNA later finds a use as an alternative *store* of genetic information—possibly a stable long-term store in spores. (*c*) the 'stable' information store also learns to replicate and then (*d*) becomes the principal genetic material—a specialist in storage and printing.

arbitrary breaking up of an original community into pieces under the influence of environmental factors such as wind and rain. A tendency to break up in a suitable way could have a selective advantage to a community by tending to increase its rate of occupation of a new territory. A piece broken off would have to be fairly large, however, for there to be a good chance that it would contain at least one of each of the genes. But the gradual organisation of the genes into 'chromosomes'—i.e. having sets of genes tied together and replicating together—would reduce the average size of a viable piece of the community. More compact 'spores' would then be possible, more easily disseminated by wind etc. But still it would be a matter of luck whether a piece of the community actually contained any or all of the 'chromosomes'—so there would be room for the evolution of mechanisms for improving the chances that a separating piece of the community would be fully genetically equipped. The result of such evolutionary processes we can see now as the elaborate procedures of mitosis in modern cells.*

Cell walls could have been evolved gradually from a vague tendency for the outer edges of a community to thicken like cold porridge to the highly sophisticated ion and molecule filters that guard the borders of the modern cell.

We can perhaps thus imagine various essential attributes of modern cellular life as having arisen gradually from the much easier gene community. In a general way we can say that the driving force behind such changes would have been the greatly increased general efficiency of the cellular mode of organisation (it would usually have been downhill, that is, on the relevant abstract landscape). Possibly this helped to 'rationalise' the intermediate heterogenetic community towards a new homogenetic solution.

But each move towards cellular organisation would have been associated with new difficulties tending to elicit new mechanisms. One of these new ideas would have been sex. In the early days, genes evolved in different communities could easily have come together by haphazard mixing of communities. With the genes all neatly parcelled out into little bags, however, this would become more difficult; yet it would remain a good idea from an evolutionary point of view for organisms to be able to try out different combinations of genes. Sex would have to be invented before cells could be much of an evolutionary success. Again this could have been simple to begin with (cells occasionally fusing) but it illustrates again that a cell, although efficient, is not the easiest form of life.

WHERE ARE THE CRYSTAL GENE SYSTEMS NOW?

You may feel that since many lowly organisms, like bacteria, are very successful there should exist somewhere examples of the still more primitive crystal genetic systems—if such ever existed. It is a mistake, however, to regard any of the organisms that are with us today as 'primitive'. No system can be called primitive that can

---

* There is still a trace of community feeling in modern cells. Mitochondria and chloroplasts have their own DNA and seem to live in a quasi-symbiotic relationship with the rest of the cell. See Goodenough and Levine, 1970.

manufacture proteins a million times faster than we can in our best-equipped laboratories. That all the organisms that we know are based on nucleic acids, and even on a common genetic code, points clearly to a common ancestor for all contemporary life forms, but it says nothing about the position of that common ancestor in the evolutionary tree (see Fig. 45(a)). That there is a single genetic system now operating means simply that other systems have become extinct. (Compare Pirie, 1959, on the extinction through natural selection of alternative biochemistries.) This is certainly a very interesting question (Why are there not *some* niches *somewhere* for the earlier systems?) but it exists independently of the question of the origin of life. Even if you think that something like nucleic acid was the first genetic material and that there was no genetic metamorphosis, the problem is still there. There must have been simpler *versions* of the system than the *version* that is now universal. (The present situation is like having all motor cars not only powered by internal combustion but by exactly the same kind of engine.) We may ask 'Where have the other *versions* gone?' If all the simpler earlier versions of the present system have been wiped out it is not reasonable to insist that truly primitive systems should have survived to this day on earth. It would, however, be reasonable to look for crystal gene systems on other planets—particularly Venus.

CONCLUSION

Figure 45(b) summarises the ideas discussed in this chapter. I have used some rather particular speculations as illustrations. Some will think that these speculations are too particular, but the main point is a general one: that the genetic material in our ultimate ancestor was very different from our own, because the requirements for a primitive genetic material are very different from the requirements for an advanced one and because a general mechanism for a radical changeover exists. Like all evolutionary processes this genetic metamorphosis would have been complicated just because it would have depended on chance events—on a cashing in on lucky accidents. It is characteristic of processes that depend in this way on chance that they take round-about routes that minimise the pieces of luck at each stage. *We* may think that it would have been sensible for life to have started in the way it meant to go on: to have used DNA right away, in spite of initial difficulties, for the sake of benefits to come. But our ultimate ancestor could not have been so far sighted.

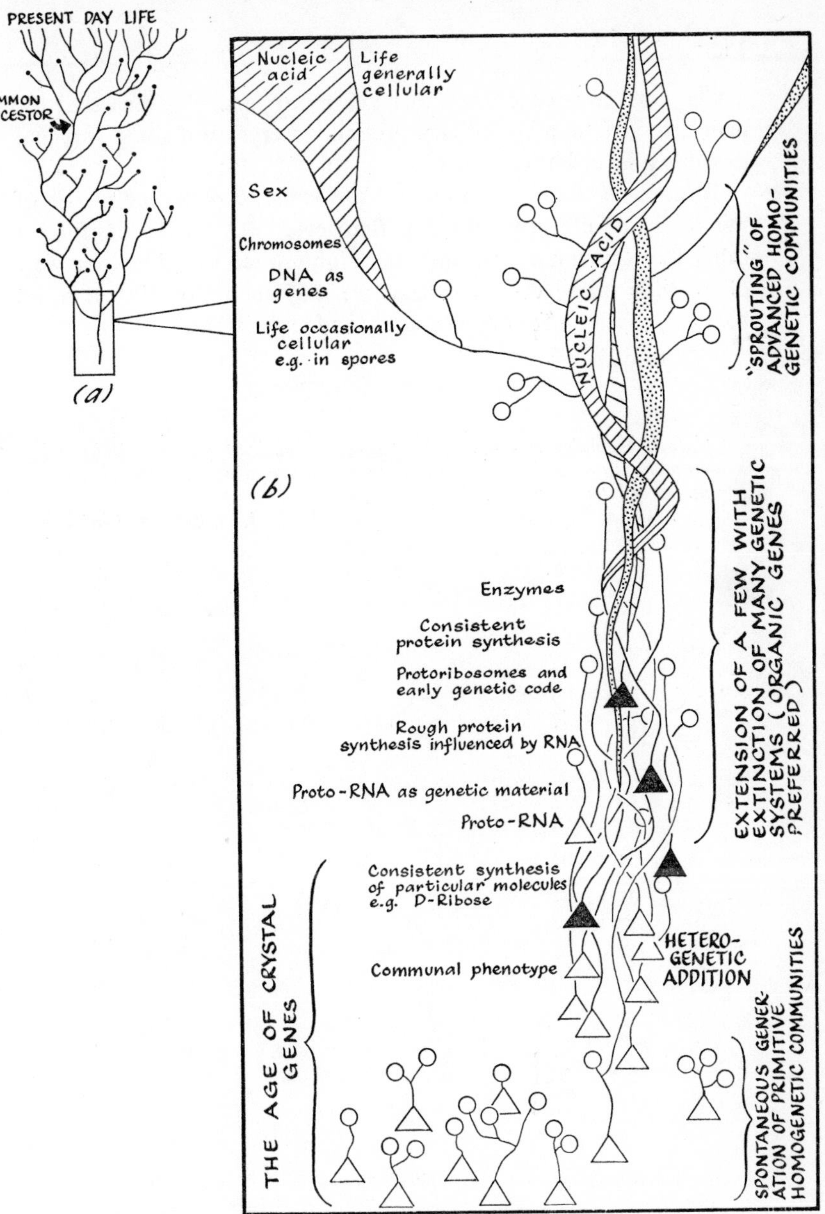

Fig. 45. That the biochemistries of all known organisms are both similar and highly sophisticated suggest a common ancestor quite high in the evolutionary tree—which might look something like (*a*). As discussed in the text, a genetic metamorphosis would take place through a number of overlapping stages. (*b*) An imaginary close-up of the lower stem of the evolutionary tree. △ represents the appearance of a new crystal genetic material, and ▲ of a new organic genetic material, ○ represents an extinction.

Chapter 7 was concerned with the more particular idea that the primitive genetic material was probably a crystal of some sort and possibly a layer-lattice silicate.

These lines of thought may seem to some to make the puzzle of the origin of life more difficult than ever. But on the contrary I think that this puzzle will only be soluble when we widen our view of possible initial events so that we may then close in on novel particular systems for detailed study and experiment.

> *'Life is beginning to cease to be a mystery and becoming practically a cryptogram, a puzzle, a code that can be broken, a working model that sooner or later can be made.'*
>
> J. D. BERNAL, 1967

References

ANFINSEN, C. B. 1963. In VOGEL, H. J. (Ed.). *Informational Macromolecules*. Academic Press, New York, pp. 153–166.

APTER, M. J. and WOLPERT, L. 1965. *Cybernetics and Development, I: Information Theory*. *J. Theoret. Biol.*, **8,** 244.

BACON, G. E. and PEASE, R. S. 1955. *Proc. Roy. Soc.*, **A230,** 359.

BALTIMORE, D. 1970. *Viral RNA-dependent DNA Polymerase*. *Nature*, **226,** 1209.

BERNAL, J. D. 1951. *The Physical Basis of Life*. Routledge and Kegen Paul, London.

BERNAL, J. D. 1959. In CLARK, F. and SYNGE, R. L. M. (Eds.). *The Origin of Life on the Earth*. Pergamon Press, London, pp. 385–399.

BERNAL, J. D. 1965. In WATERMAN, T. H. and MOROWITZ, H. J. (Eds.). *Theoretical and Mathematical Biology*. Blaisdell Publ. Co. New York.

BERNAL, J. D. 1966. In WOLSTENHOLME, G. E. W. and O'CONNOR, M. (Eds.). *Principles of Biomolecular Organisation*. Churchill, London, pp. 1–6.

BERNAL, J. D. 1967. *The Origin of Life*. Weidenfeld and Nicolson, London.

BLUM, M. S. 1955. *Time's Arrow and Evolution*. Harper, New York.

BRAGG, L. 1968. *X-Ray Crystallography*. *Scientific American*, **219,** No. 1, 58–70.

BROWN, G. and WEIR, A. H. 1963. In ROSENQVIST, I. T. and GRAFF-PETERSON, P. (Eds.). *International Clay Conference*. Pergamon, Oxford, p. 28.

CAILLÈRE, S., OBERLIN, A. and HÉNIN, S. 1954. *Etude au microscope électronique de quelques silicates phylliteux obtenus par synthèse à basse température*. *Clay Min. Bull.*, **2,** No. 12, 146–157.

CAIRNS-SMITH, A. G. 1961. *Tautomerism in the Solid State, I: Thermochromism of Heterocyclic Phenols*. *J. Chem. Soc.*, 182–188.

CAIRNS-SMITH, A. G. 1966. *The Origin of Life and the Nature of the Primitive Gene*. *J. Theoret. Biol.*, **10,** 53–88.

CAIRNS-SMITH, A. G. 1968. *An Approach to a Blueprint for a Primitive Organism*. In WADDINGTON, C. H. (Ed.). *Towards a Theoretical Biology Vol. I: Prolegomena*. Edinburgh University Press, Edinburgh, pp. 57–66.

CAIRNS-SMITH, A. G. and PETTIGREW, R. 1969. *Synthesis of Nylon-like Oligoamides*. *J. Chem. Soc.* (*C*), 1606.

CALVIN, M. 1965. *Chemical Evolution*. *Proc. Roy. Soc.*, **288A,** 441.

CALVIN, M. 1969. *Chemical Evolution. Molecular Evolution towards the*

Origin of Living Things on the Earth and Elsewhere. Oxford University Press, London.

CHANGEAUX, J.-P. 1965. *The Control of Biochemical Reactions. Scientific American*, **212,** No. 4, 36–45.

CHEDD, G. 1970. *Doubts about how DNA is copied. New Scientist*, **46,** No. 703, p. 426.

CHIPMAN, D. M. and SHARON, N. 1969. *Mechanism of Lysozyme Action. Science*, **165,** 454.

CLOUD, P. E. 1968. *Atmospheric and Hydrospheric Evolution on the Primitive Earth. Science*, **160,** 729–736.

COOK, R. A. and KOSHLAND, D. E. 1969. *Specificity in the Assembly of Multi-Subunit Proteins. Proc. Natl. Acad. Sci.*, **64,** 247.

DANCOFF, S. M. and QUASTLER, H. 1953. In QUASTLER, H. (Ed.). *Information Theory in Biology*. Urbana, Illinois, p. 263.

DOTY, P. and YANG, J. T. 1956. *J. Amer. Chem. Soc.*, **78,** 498.

DOTY, P., WADA, A., YANG, J. T. and BLOUT, E. R. 1957. *J. Polymer Sci.*, **23,** 851.

EDDINGTON, A. S. 1935. New Pathways in Science. Cambridge University Press, Cambridge, p. 221.

EGLINTON, G. and CALVIN, M. 1967. *Chemical Fossils. Scientific American*, **216,** No. 1, 32–43.

EITEL, W. 1966. *Silicate Science vol. 4: Hydrothermal Silicate Systems.* Academic Press, New York, p. 363.

FIRSOFF, V. A. 1967. *C.S.P. No. 2: Life, Mind and Galaxies*. Oliver and Boyd, Edinburgh.

FISCHER, E. 1894. *Ber.*, **27,** 298.

FISCHER, E. W. 1957. *Z. Naturf.*, **A12,** 753.

FLORENTIEV, V. L. and IVANOV, V. I. 1970. *RNA Polymerase: Two Step Mechanism with Overlapping Steps. Nature*, **228,** 519.

FOX, S. W. (Ed.). 1965. *The Origins of Prebiological Systems*. Academic Press, New York.

FRANK, F. C. 1949. *Discuss. Faraday Soc.*, **5,** 54.

GAMOW, G. 1961. *The Creation of the Universe*. Macmillan, London.

GATINEAU, L. 1964. *Bull. Soc. fr. Miner, Crystallogr.*, **87,** 321.

GIBSON, G. E. and GIAUQUE, W. F. 1923. *J. Amer. Chem. Soc.*, **45,** 93.

GOODENOUGH, U. W. and LEVINE, R. P. 1970. *The Genetic Activity of Mitochondria and Chloroplasts. Scientific American*, **223,** No. 5, 22.

GRANQUIST, W. T. and POLLACK, S. S. 1960. *Clays, Clay Minerals*, **8,** 150.

HALDANE, J. B. S. 1929. *The origin of life. Rationalist Annual*, **3.** (Reprinted in Bernal, 1967.)

HALDANE, J. B. S. 1930. *Enzymes*. Longmans, London, p. 182.

HALDANE, J. B. S. 1954. *The Origins of Life. New Biology*, **16,** 12–27.

HALDANE, J. B. S. 1965. *Data Needed for a Blueprint of the First Organism.* In FOX, S. W. (Ed.). *The Origins of Prebiological Systems*. Academic Press, New York, pp. 11–18.

HEATH, T. L. 1897. *The Works of Archimedes*. Cambridge University Press, Cambridge, p. 232.

HÉNIN, S. and ROBICHET, O. 1954. *A Study of the Synthesis of Clay Minerals. Clay Min.Bull.*, **2,** No. 11, 110–115.

HENZE, H. R. and BLAIR, C. M. 1931. *Number of Isomeric Hydrocarbons of the Methane Series. J. Amer. Chem. Soc.*, **53,** 3077.

HINSHELWOOD, C. 1951. *The Structure of Physical Chemistry.* Clarendon Press, Oxford, p. 83.

HINTON, H. E. and BLUM, M. S. 1965. *Suspended Animation and the Origin of Life. New Scientist.*, **28,** 270–271.

HOROWITZ, N. H. 1959. *On Defining Life.* In CLARK, F. and SYNGE, R. L. M. (Eds.). *The Origin of Life on the Earth.* Pergamon Press, London, pp. 106–107.

HUXLEY, J. 1963. *Evolution: The Modern Synthesis.* Allen and Unwin, London, Introduction.

KELLER, A. 1957. *Phil. Mag.*, **2,** 1171.

KELLER, A. 1962. *Polymer*, **3,** 393.

KELLER, A. 1964. *Kolloid Z.*, **197,** 98.

KENDREW, J. C. 1961. *The Three Dimensional Structure of a Protein Molecule. Scientific American*, **205,** No. 6, 96–111.

KENDREW, J. C. and WATSON, H. C. 1966. In WOLSTENHOLME, G. E. W. and O'CONNOR, M. (Eds.). *Principles of Biomolecular Organisation.* Churchill, London, pp. 86–100.

KITAIGORODSKII, A. I. 1961. *Organic Chemical Crystallography.* Translated from the Russian. Consultants' Bureau, New York.

KITAIGORODSKII, A. I. 1970. *General View on Molecular Packing. Advances in Structure Research by Diffraction Methods*, **3,** 173.

KOLLER, P. C. 1968. *C.S.P. No. 30: Chromosomes and Genes.* Oliver and Boyd, Edinburgh.

KURLAND, C. G. 1970. *Ribosome Structure and Function Emergent. Science*, **169,** 1171.

LAEMMLI, U. K. 1970. *Cleavage of Structural Proteins during the Assembly of the Head of Bacteriophage T.4. Nature*, **227,** 680.

LAMBERT, J. B. 1970. *The Shapes of Organic Molecules. Scientific American*, **222,** No. 1, 58.

LIPMANN, F. 1969. *Polypeptide Chain Elongation in Protein Biosynthesis. Science*, **164,** 1024.

MARQUAND, J. 1968. *C.S.P. No. 17: Life: its Nature, Origins and Distribution.* Oliver and Boyd, Edinburgh.

MARSH, R. E. and DONOHUE, J. 1967. *Crystal Structure Studies of Amino-acids and Peptides. Advanc. Protein Chem.*, **22,** 235.

MAYNARD SMITH, J. 1962. In GOOD, I. J. (Ed.). *The Scientist Speculates.* Heinemann, London.

MAYNARD SMITH, J. 1966. *The Theory of Evolution.* Penguin Books, Harmondsworth, p. 96.

MILLER, S. L. 1953. *Science*, **117,** p. 528.

MONOD, J., CHANGEAUX, J.-P. and JACOB, F. 1963. *Allosteric Proteins and Cellular Control Systems. J. Molecular Biol.*, **6,** 306–29.

MOORE, E. F. 1964. *Mathematics in the Biological Sciences. Scientific American*, **211,** No. 3, 154.

MOROWITZ, H. J. 1966. *The Minimum Size of Cells.* In WOLSTENHOLME, G. E. W. and O'CONNOR, M. (Eds.). *Principles of Biomolecular Organisation.* Churchill, London, pp. 446–462.

MORTLAND, M. M. and PINNAVAIA, T. J. 1971. *Formation of Copper (II) Arene Complexes on the Interlamelar Surfaces of Montmorillonite. Nature Phys. Sci.*, **229**, 75.

NEURATH, H., WALSH, K. A. and WINTER, W. P. 1967. *Evolution of Structure and Function in Proteases. Science*, **158**, 1638–1644.

NOMURA, M. 1969. *Ribosomes. Scientific American*, **221**, No. 4, 28.

NORTH, A. C. T. 1966. *The Structure of Lysozyme. Science Journal*, **2**, No. 11, 55–63.

OPARIN, A. I. 1924. *Proiskhozhdenie zhizni*, Moscow. (Reprinted in BERNAL, 1967.)

OPARIN, A. I. 1957. *The Origin of Life on the Earth.* Oliver and Boyd, Edinburgh.

PAECHT-HOROWITZ, M., BERGER, J. and KATCHALSKY, A. 1970. *Prebiotic Synthesis of Polypeptides by Heterogeneous Polycondensation of Amino Acid Adenylates. Nature*, **228**, 636.

PASK, G. 1961. *An Approach to Cybernetics.* Hutchinson, London.

PAULING, L. 1960. *The Nature of the Chemical Bond.* Cornell University Press, New York, 3rd Edition.

PEDRO, G. 1962. *Genèse des Minéraux Argilleux par Lessivage des Roches Cristallines au Laboratoire.* In *Genèse et Synthèse des Argiles.* Centre National de la Recherche Scientifique, Paris, pp. 99–107.

PERUTZ, M. F. 1964. *The Hemoglobin Molecule. Scientific American*, **210**, No. 11, 64–76.

PHILLIPS, D. C. 1966. *The Three Dimensional Structure of an Enzyme Molecule. Scientific American*, **215**, No. 5, 78.

PIRIE, N. W. 1959. In CLARK, F. and SYNGE, R. L. M. (Eds.). *The Origin of Life on the Earth.* Pergamon, London, pp. 76–83.

PITTENDRIGH, C. S. 1958. In ROE, A. and SIMPSON, G. G. (Eds.). *Behaviour and Evolution.* Yale University Press, New Haven, Conn.

RADOSLOVICH, E. W. and JONES, J. B. 1961. *Clay Min. Bull.*, **4**, No. 26, 318–322.

REX, R. W. 1966. *Authigenic Kaolinite and Mica as Evidence for Phase Equilibria at Low Temperatures.* In BRADLEY, W. F. and BAILEY, S. W. (Eds.). *Clays and Clay Minerals, 13th Conference.* Pergamon, Oxford, pp. 95–104.

RILEY, P. A. 1970. *A Suggested Mechanism for DNA Transcription. Nature*, **228**, 522.

SCHRODINGER, E. 1944. *What is Life?* Cambridge University Press, Cambridge.

SHERRINGTON, C. 1940. *Man on His Nature (1937 Gifford Lectures).* Cambridge University Press, London, p. 88.

SILLEN, L. G. 1967. *How Sea Water and Air got their Present Compositions. Chemistry in Britain*, **3**, 291

SIMPSON, G. G. 1950. *The Meaning of Evolution.* Oxford University Press, London, p. 15.

TATUM, G. W. and BEADLE, G. 1941. *Proc. Nat. Acad. Sci.*, **27**, 499–506 and 1945. *Am. Naturalist*, **79**, 304.

TAUB, A. H. (Ed.) 1963. *John von Neumann Collected Works, vol. 5.* Pergamon, Oxford, 288.

TEMIN, H. M. and MIZUTANI, S. 1970. *RNA-dependent DNA Polymerase in Virions of Rous Sarcoma Virus. Nature*, **226,** 1211.

TILL, P. H. Jr. 1957. *J. Polymer Sci.*, **24,** 301.

TOULMIN, S. and GOODFIELD, J. 1965. *The Architecture of Matter.* Penguin Books, Harmondsworth, p. 92.

VAN OLPHEN, H. I. 1963. *An Introduction to Clay Colloid Chemistry.* Interscience, New York, p. 170.

WADDINGTON, C. H. 1957. *The Strategy of the Genes: a Discussion of some Aspects of Theoretical Biology.* Allen and Unwin, London.

WADDINGTON, C. H. (Ed.) 1968. *Towards a Theoretical Biology 1: Prolegomena.* Edinburgh University Press, Edinburgh and Aldine Press, Chicago.

WATSON, J. D. and CRICK, F. H. C. 1953. *Nature*, **171,** 964.

WEIR, A. H., NIXON, H. L. and WOODS, R. D. 1962. Measurement of thickness of dispersed clay flakes with the electron microscope. In *Clays and Clay Minerals, 9th Conference* (E. Ingerson, editor), pp. 419–423. Pergamon Press, Oxford.

WEISS, A. 1963. *Mica-type Layer Silicates with Alkylamonium Ions.* In SWINEFORD, A. (Ed.). *Clays and Clay Minerals, 10th Conference.* Pergamon, Oxford, pp. 191–224.

WEY, R. and SIFFERT, B. 1962. *Réactions de la Silice Monomoléculaire en Solution avec les Ions Al^{3+} et Mg^{2+}.* In *Genèse et Synthèse des Argiles.* Centre National de la Recherche Scientifique, Paris, pp. 11–23.

WOLSTENHOLME, G. E. W. and O'CONNOR, M. (Eds.) 1966. *Principles of Biomolecular Organisation.* Churchill, London.

WOOD, W. B. and EDGAR, R. S. 1967. *Building a Bacterial Virus. Scientific American*, **217,** No. 1, 60–75.

ZIMM, B. H. and BRAGG, J. K. 1959. *J. Chem. Phys.*, **31,** 526.

ZUCKERKANDL, E. 1965. *The Evolution of Hemoglobin. Scientific American*, **212,** No. 5, 110–118.

Index